LIGHT MICROSCOPY

LIGHT MICROSCOPY
ESSENTIAL DATA

C.P. Rubbi

Department of Biology, University of Essex, Colchester, UK

JOHN WILEY & SONS
Chichester · New York · Brisbane · Toronto · Singapore

Published in association with BIOS Scientific Publishers Limited

©BIOS Scientific Publishers Limited, 1994. Published by John Wiley & Sons Ltd, Baffins Lane, Chichester, West Sussex PO19 1UD, UK, in association with BIOS Scientific Publishers Ltd, St Thomas House, Becket Street, Oxford OX1 1SJ, UK.

British Library Cataloguing in Publication Data
A catalogue record for this book is available from the British Library.

ISBN 0 471 94270 7

Library of Congress Cataloging in Publication Data
Rubbi, C.
  Light microscopy: essential data/C. Rubbi.
    p. cm.—(Essential data series)
    'Published in association with BIOS Scientific Publishers Limited.'
    Includes bibliographical references and index.
    ISBN 0 471 94270 7
    1. Microscopy—Laboratory manuals. I. Title. II. Series.
QH205.2.R83 1994
578'.4——dc20   94-9543
                                                              CIP

Typeset by Marksbury Typesetting Ltd, Bath, UK
Printed and bound in UK by H. Charlesworth & Co. Ltd, Huddersfield, UK

CONTENTS

Abbreviations   x
Preface   xiii

1. Basic aspects of light microscopy   1
  Choice of microscopic method   1
  Image formation by a thin lens   1

  Figures and tables
  Flow-chart for the selection of microscopic
    observation method   2
  Comparison of observation methods for
    cells fluorescently labeled for surface antigen   3
  Properties of thin lenses and apertures   4
  Fundamental equations for lenses and
    microscopes   5
  Image formation by a thin lens   6

2. Types of microscopy   7
  The compound microscope and Köhler
    illumination (bright field observation)   7
  Phase contrast   10
  Dark field   11
  Polarization microscopy   11
  Differential interference contrast microscopy
    (Nomarski optics)   16
  Interferometer microscope   18
  Hoffman modulation contrast microscopy   18
  Interference reflection microscopy   21
  Fluorescence microscopy   21
  Confocal microscopy   23

  Figures and tables
  Components of a compound microscope
    using Köhler illumination   8

Principal components of a light microscope 12
A phase contrast microscope 14
Image formation in a dark field microscope 15
Image formation in a polarization microscope 15
Components and image formation in a
  differential interference contrast microscope 17
Components and image formation in a
  Hoffman modulation contrast microscope 19
Interfering beams in interference reflection
  microscopy 21
Epiillumination assemblies for interference
  reflection and fluorescence microscopes 22
Confocal microscopy 24
Role of each component of a compound
  microscope 26
Procedure for setting Köhler illumination 27
Compensators for polarization microscopy 28
Features and components of the polarization
  microscope 28
Image formation in a DIC microscope 29

Set-up for Hoffman modulation contrast 30
Equipment requirements for fluorescence
  microscopy 31

3. Filters and mirrors for microscopy 32
Interference filters and chromatic beam splitters 32
Choice of filter sets for fluorescence applications 34

Figure and tables
Interference filters 33
Types of interference filters 35
Selection of filter sets for fluorescein detection 36

4. Aberrations 37
Figure and table
Aberrations 38
Lens aberrations 41

5. Light sources 43
Lamps 43
Lasers 44

Tables
Comparison of incandescent and arc light
   sources   45
Characteristics of some common lasers   45

6. Lens nomenclature   **46**

7. Measurement methods   **48**
Size measurement   48
Counting   48

Figure and Tables
Scheme of an improved Neubauer chamber   49
Linear measurement methods   50
Typical sizes of cells and organelles   51

8. Photomicrography   **52**

Table
Characteristics of some Kodak films for use
   in microscopy   53

9. Fixation   **54**

Table
Properties of some common chemical fixatives   55

10. Embedding and cutting   **56**
Dehydration   56
Clearing   56
Decalcification   57
Embedding   57
Cutting   58

Tables
Summary of decalcifying agents   58
Some commonly used embedding media   59
Types of microtomes   59

11. Staining   **60**
Non-fluorescent stains   60
Fluorescent stains   61
Fluorescent analogs of biomolecules   62

Contents

Fluorescent physiological indicators and tracers 63

Figures and tables
Spectral variations within the BODIPY
family of fluorochromes 63
Spectra of Indo-l at increasing concentrations
of Ca^{2+} 64
Properties of some common dyes 66
Biochemical phenomena used for specific
staining 68
Enzyme–substrate combinations for
immunohistochemistry 69
Substituents and resulting spectra for
BODIPY fluorophores 69
Reactive groups for the synthesis of
fluorescent analogs of biomolecules 69
Some commonly used fluorophores 70
Examples of ratiometric probes 71
Fluorescent viability tests 71
Examples of organelle-specific probes 72

**12. Specific delivery of stains: immuno-
histochemistry and immunofluorescence 73**
General procedures 73
(Strept)avidin–biotin methods 74
Enzyme–anti-enzyme methods 74

Tables
Properties of immunoglobulins and
immunoglobulin fractions for use in
immunohistology 76
Protocol for immunofluorescent labeling of
surface antigens 77
The peroxidase–anti-peroxidase method 78

13. DNA probes for *in situ* hybridization 79

14. Anti-fading agents and specimen mounting 80

15. Suppliers of equipment and chemicals 81
Microscopes and ancillary equipment 81
Chemicals, stains and antibodies 83

Further reading 86
General texts 86
Articles and manuals 87

Glossary 88

Appendix 90
Care and maintenance of the light microscope 90

Index 92

Contents

ABBREVIATIONS

APAAP	alkaline phosphatase–anti-alkaline phosphatase
BODIPY	4,4-difluoro-4-bora-3α,4α-diaza-s-indacene
d_z	axial resolution for reflection of a plane
DABCO	1,4-diazabicyclo[2.2.2]octane
DASPEI	2-(4-dimethylaminostyryl)-*N*-ethylpyridinium iodide
df	depth of field
2-Di-l-ASP (DASPMI)	2-(4-dimethylaminostyryl)-*N*-methylpyridinium iodide
4-Di-l-ASP (DASPMI)	2-(4-dimethylaminostyryl)-*N*-methylpyridinium iodide
4-Di-2-ASP (DASPMI)	2-(4-dimethylaminostyryl)-*N*-methylpyridinium iodide
DIC	differential interference contrast
DiOC$_5$(3)	3,3′-dipentyloxacarbocyanine iodide
DiOC$_6$(3)	3,3′-dihexyloxacarbocyanine iodide
DiOC$_7$(3)	3,3′-diheptyloxacarbocyanine iodide
EDCK	*N,N*′-bis-(2-ethyl-1,3-dioxolane) kryptocyanine
FITC	fluorescein isothiocyanate
FMI-43	*N*-(3-triethylammoniumpropyl)-4-(*p*-dibutylaminostyryl)pyridinium dibromide
HBO	high-pressure mercury arc lamp
IMS	industrial methylated spirit
IR	infra-red light
LSCM	laser scanning confocal microscope
n	refractive index
$n_\parallel$	refractive index for light parallel to optical axis
$n_\perp$	refractive index for light perpendicular to optical axis

NA	numerical aperture		(4-*p*-diethylaminophenyl)-
NBD ceramide	6-((*N*-(7-nitrobenz-2-oxa-1,3-diazol-4-yl) amino)caproyl)sphingosine	SSCM	butadienyl)pyridinium dibromide stage scanning confocal microscope
NBD sphingo-myelin	*N*-(6-(7-nitrobenz-2-oxa-1,3-diazol-4-yl) amino)hexanoyl)sphingosylphosphocholine	UV XBO	ultra-violet light high-pressure xenon arc lamp
PAP	peroxidase–anti-peroxidase	Δ	optical path difference
PMT	photomultiplier	$\Delta\phi$	phase difference, phase shift
R	lateral resolution (Rayleigh's criterion)	ε	extinction coefficient
RH 414	*N*-(3-triethylammoniumpropyl)-4-	λ	wavelength

PREFACE

This book provides a summary of microscopic data and observation procedures for biological specimens.

The first seven chapters describe the fundamentals and operation of the most common microscopes, as well as selected data on related topics such as lenses, lamps, mirrors and aberrations. In each subject, theoretical treatments have been kept to the minimum necessary for a reasonable level of comprehension.

The later chapters provide important data on specimen preparation methods, especially specific labeling procedures such as immunohistochemical and immunofluorescent methods. Care has been taken to include data on a wide variety of fluorochrome molecules and their applications, including caged compounds, ion and pH indicators, etc. I am grateful to Dr Richard D. Jurd for his contribution to part of Chapter 10.

This book is designed as a quick reference for the operation of all types of microscopes as well as the most common methodological procedures. A detailed treatment of each topic (especially of theoretical aspects) will require the reader to consult one of the more specialized books which are given in the Further Reading section.

C.P. Rubbi

Chapter 1 BASIC ASPECTS OF LIGHT MICROSCOPY

1 Choice of microscopic method

The choice of microscopic procedure depends on a number of variables such as the type of specimen, the staining method and the resolution required, as well as the type of microscope that is available. The flow-chart depicted in *Figure 1* is a guide to the selection of the appropriate microscopic observation method, with respect to the specimen characteristics and the features to be observed. *Figure 2* shows a group of cells photographed under different observation conditions.

2 Image formation by a thin lens

When an object is placed at a distance p from a thin lens, an image is formed at a distance q. These distances are related by equation 1 in *Table 1*, which defines the focal distance of the lens and, thus, its two foci. This treatment assumes that objects are placed at or close to the axis of the lens (paraxial). The planes normal to the optical axis that contain the foci are called front focal plane if located on the object side and back focal plane if otherwise. For analyzing light paths, a general ray-tracing rule for thin lenses can be deduced:

(1) rays passing through the center of the lens continue undeviated;

(2) rays approaching the lens parallel to the optical axis will pass through the opposite focus (conversely, rays originated in a focus will continue parallel to the optic axis after passing the lens).

General aspects of image formation by a thin lens, as well as roles of apertures, that are relevant to the following sections are summarized in *Table 2* and corresponding schemes are shown in *Figure 3*.

1

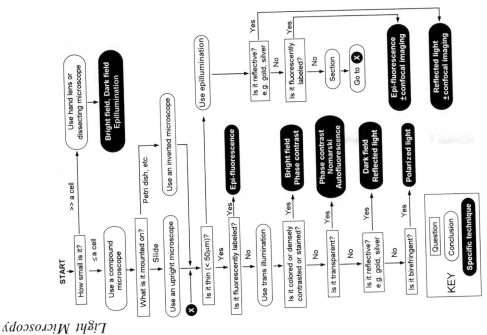

Figure 1. Flow-chart for the selection of microscopic observation method. Reproduced from Rawlins (1992), *Light Microscopy*, Figure 1.1, BIOS Scientific Publishers.

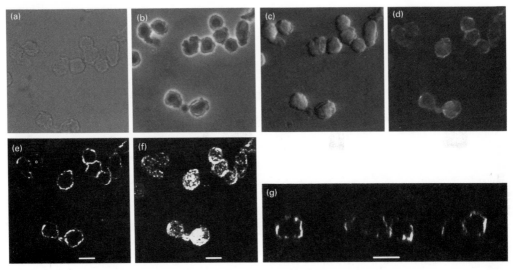

Figure 2. Comparison of observation methods for a group of cells fluorescently labeled for a surface antigen. (a) Bright field, (b) phase contrast, (c) Hoffman modulation contrast, (d) conventional fluorescence, (e) confocal *x–y* optical section, (f) confocal extended focus image, (g) confocal vertical section. Magnification: × 51. Bar = 10 μm.

3   *Basic Aspects of Light Microscopy*

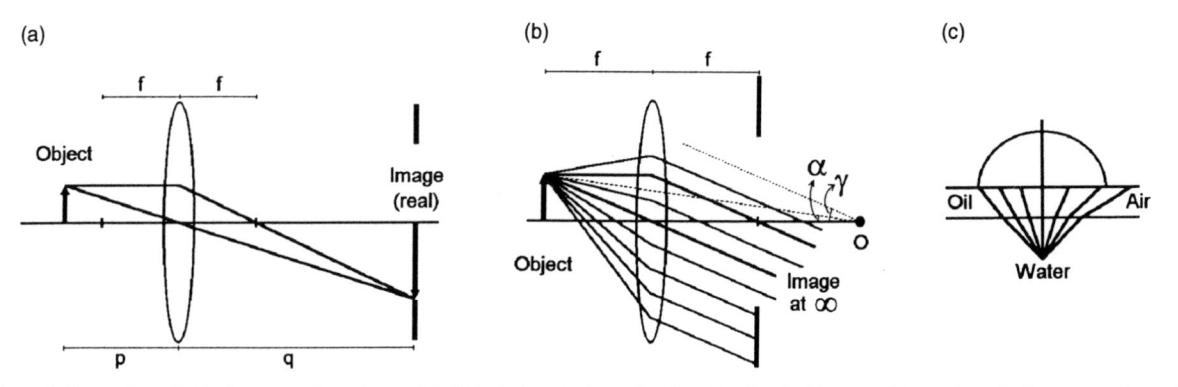

Figure 3. Properties of thin lenses and apertures. (a) Object placed at greater than the focal distance, with aperture in the image plane; (b) object placed in a focus, with aperture in the other focus; (c) effect of immersion medium.

Table 1. Fundamental equations for lenses and microscopes

Magnitude	Equation		Comments
Focal distance	$1/p + 1/q = 1/f$	(1)	p, distance from object to lens; q, distance from image to lens; f, focal distance
Magnification	$M = \gamma/\alpha$	(2a)	Magnification in the case of *Figure 3b*.
	$M = q/p$	(2b)	Magnification in the case of *Figure 3a*
Resolution (lateral)	$R = 0.61 \times \lambda/NA_{objective}$	(3a)	R, lateral resolution (Rayleigh's criterion), as minimum distance between the centers of two resolvable points for either self-luminous points or incoherently illuminated; NA, numerical aperture; λ, wavelength. For confocal microscopy R can be $\sqrt{2}$ times smaller
	$R = 1.22\lambda/(NA_{objective} + NA_{condenser})$	(3b)	As equation 3a, for $NA_{condenser} \neq NA_{objective}$
	$R = 1.22\lambda/(NA_{objective})$	(3c)	As equation 3a, for coherent illumination such as $NA_{condenser} = 0$ (non-oblique rays) or laser light. For fluorescence (self-luminous points), equation 3a applies, even in the case of laser illumination
Depth of field	$df = 2\lambda/(n \times \sin^2\theta)$	(4)	df, depth of field; θ, half angle subtended to the illuminated point ($\sin\theta = NA/n$)
Resolution (axial–confocal)	$d_z = 0.45\lambda/(n(1-\cos\theta))$	(5)	d_z, axial resolution for reflection of a plane in a confocal microscope (width at half maximum intensity)
Image brightness	$E = \pi L_o(NA/(nM))^2$	(6)	E, image illumination; L_o, object's luminance; NA, objective's numerical aperture; M, objective's magnification; n, refractive index

Basic Aspects of Light Microscopy

Table 2. Image formation by a thin lens

Condition	Result[a]	Example
Object placed at a distance from the lens that is longer than the focal distance	Real image formed beyond opposite focal plane, as in *Figure 3a*. Distances related by equation 1. Magnification defined as in equation 2b	Primary image of the specimen formed by the objective lens
Object placed at the focal distance	Image formed at infinity: rays exit the lens as a parallel beam, as in *Figure 3b*. Magnification defined as in equation 2a	Image formed by the eyepiece (from the primary image that is placed at its focal plane) (*Figure 1a*, Chapter 2)
Stop in the image/object plane	Scheme in *Figure 3a*. The field of view is limited by the size of the stop. No limitations in resolution or image brightness. The same effect is obtained by locating the stop in any plane conjugate with the image/object plane	Field iris (*Figure 1a*, Chapter 2)
Stop in a focal plane (not in image plane)	Scheme in *Figure 3b*. Affects resolution and depth of field (equations 3a, b, c and 4) by limiting NA. Also affects image brightness (equation 6). Does not affect image size	Condenser iris (*Figure 1b*, Chapter 2)
Immersion medium	Increases lens light gathering power by allowing rays at higher angles to be collected. Higher NA can be achieved. Comparative scheme in *Figure 3c*	Oil immersion objectives

[a]Equations as in *Table 1*.

Chapter 2 TYPES OF MICROSCOPY

1 The compound microscope and Köhler illumination (bright field observation)

The compound microscope combines two lens systems: the objective, which forms a real image, and the eyepiece, which forms an image at infinity which can be viewed by an observer. A schematic representation of the assembly is depicted in *Figure 1a*. The total magnification is the product of the magnifications of the two lenses. The scheme also shows two additional lenses, the condenser and the collector, that provide a convenient illumination.

Currently, the assembly is illuminated following the procedure introduced by Köhler that simultaneously creates an evenly illuminated field of view (parallel rays), while illuminating the specimen with a cone of light as wide as possible. Two series of planes are generated (conjugate planes) one containing successive images of the specimen (image-forming series – *Figure 1a*) and the other containing successive images of the lamp filament (illuminating series – *Figure 1b*). The locations of each plane series are:

Illuminating rays	*Image-forming rays*
Lamp filament	Field diaphragm
Condenser aperture (front focal plane)	Specimen plane
Objective back focal plane	Primary image (front focal plane of eyepiece)
Exit pupil of the eyepiece (observer's pupil)	Retina of the eye

Note that at the specimen plane, illuminating rays are parallel, while image-forming rays form a wide cone. Elements in the

(a)

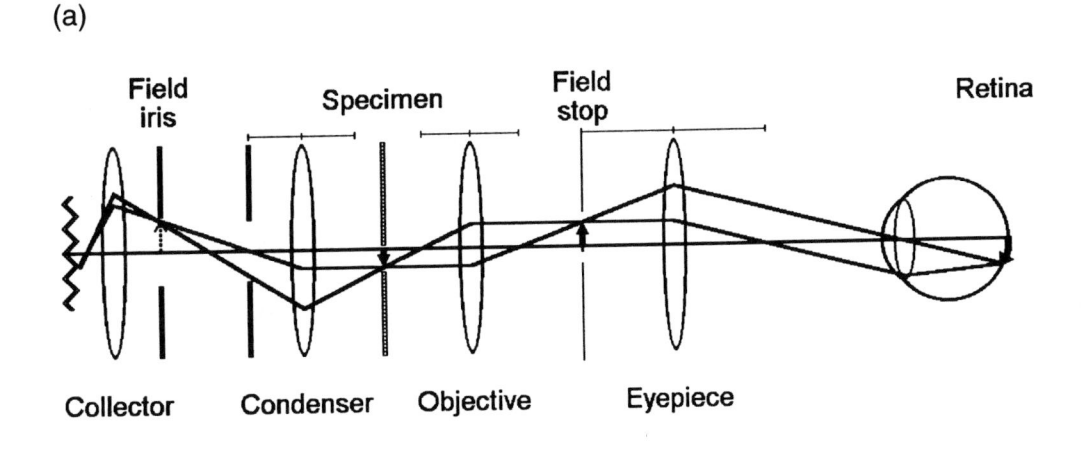

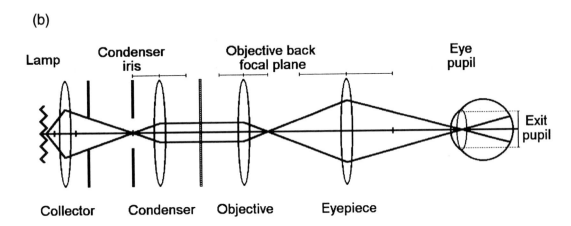

(b)

Figure 1. Components of a compound microscope using Köhler illumination. Bars over lenses indicate focal distances. (a) Conjugate planes for image-forming rays; (b) conjugate planes for illuminating rays.

Types of Microscopy

image-forming conjugate planes are viewed when observing through the eyepiece. Elements in the illuminating conjugate planes are viewed by focusing on the objective back focal plane with the aid of a phase telescope or Bertrand lens.

Figure 2 shows a scheme of a light microscope, indicating its major components. The roles of each component are summarized in *Table 1*. The procedure for setting up Köhler illumination is summarized in *Table 2*. Unstained cells viewed in bright field are shown in *Figure 2a* of Chapter 1.

2 Phase contrast

In the phase contrast method, developed by Zernike, light passing through a thin transparent specimen is considered to be split into two parts:

1. Background light travelling undiffracted and traversing the specimen as parallel beams (Köhler illumination is assumed);
2. Waves generated by diffraction on interacting with the specimen suffer a $\lambda/4$ retardation with respect to the undiffracted (background) light.

When the image is formed in bright field conditions, the phase difference ($\lambda/4$) between background and diffracted light will not be large enough for either destructive or constructive interference and hardly any specimen detail will be observed. However, if the undiffracted light is advanced by $\lambda/4$, the total phase difference at the image plane will be $\lambda/2$, thus resulting in destructive interference and producing dark specimen details on a bright background. Separate treatment of diffracted and undiffracted light can be performed by placing an annular aperture in the front focal plane of the condenser and a phase ring in the back focal plane of the objective, as depicted in *Figure 3a*. The phase ring consists of a glass slide with an annular $\lambda/4$ depression (*Figure 3b*). When the system is adequately set up, only undiffracted light will pass through the $\lambda/4$ depression, thus suffering a $\lambda/4$ phase advance because it travels further through a medium (air) of lower refractive index than glass. Light diffracted from the specimen will pass outside of the $\lambda/4$ depression, originating a total $\lambda/2$ phase retardation with respect to background light. The resulting destructive interference will cause object details to appear darker and the effect is called *positive phase contrast*. The $\lambda/4$ depression has a light-absorbing mask which attenuates background light, thus increasing the contrast. If an annular $\lambda/4$ elevation is used (*Figure 3b*), the interference is constructive and the system will produce *negative phase contrast* (refractive objects brighter than background). Each phase contrast objective lens has a built-in phase plate, thus a specific condenser annular aperture is required for each one. The set-up for phase contrast is described as part of the Köhler illumination set-up (*Table 2*). Unstained cells viewed under phase contrast (positive low) are shown in *Figure 2b* of Chapter 1.

3 Dark field

In dark field illumination, light from the condenser reaches the specimen at an angle that cannot be accepted by the objective's aperture (its NA). As a result, only highly diffracting or scattering structures can be observed on a dark background. Such a condenser, which uses mirrors instead of lenses, is depicted in *Figure 4*. It can be attached to a normal objective or to one with a variable, aperture-limiting iris. Oil immersion condensers are often used. In this method, alignment is critical and the procedure (similar to the phase-contrast set-up described in *Table 2*) has to be performed carefully.

4 Polarization microscopy

A polarizing microscope includes two polaroids, one between the lamp and the condenser (named the polarizer) and the other between the objective and the eyepiece

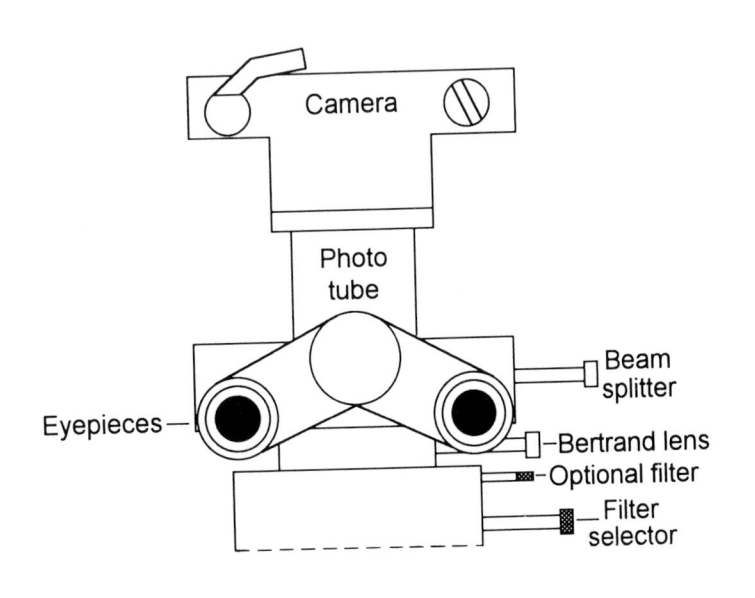

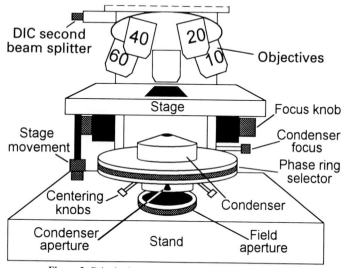

Figure 2. Principal components of a light microscope.

Types of Microscopy

(a)

(b)

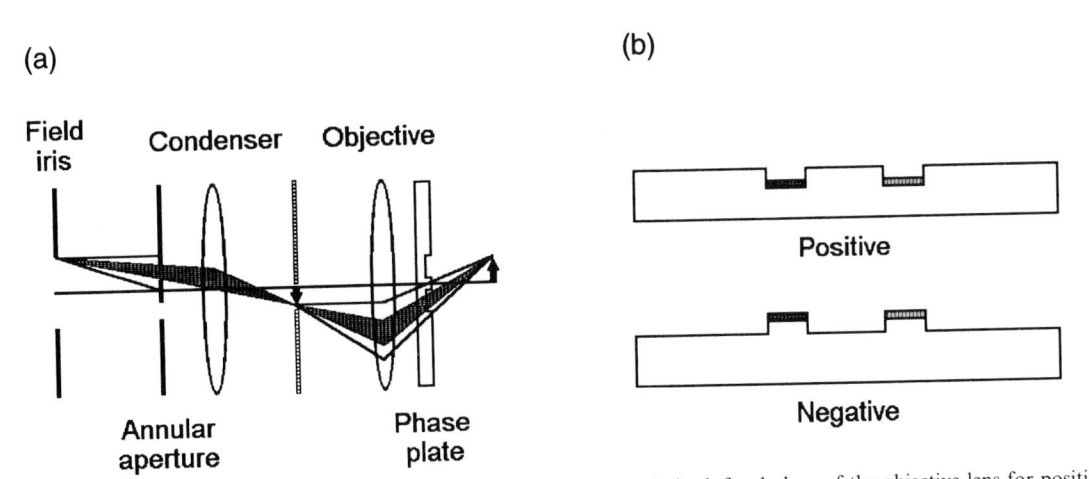

Figure 3. A phase contrast microscope. (a) Image formation; (b) phase plates in back focal plane of the objective lens for positive and negative contrast (shadowed areas indicate the light-absorbing masks – see text).

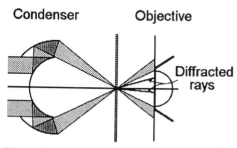

Figure 4. Image formation in a dark field microscope.

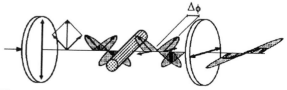

Figure 5. Image formation in a polarization microscope.

(named the analyzer). The analyzer is crossed with respect to the polarizer and thus no light can reach the observer. If a specimen containing birefringent structures is focused (*Figure 5*), these will split the polarized incident light into two components, one polarized along its fast axis and the other along the slow one. As a result, the exiting components will have a phase difference ($\Delta\phi$ in *Figure 5*) with respect to each other, but they cannot interfere because they are polarized at different angles. The phase difference is related to the birefringence by:

$$\Delta\phi = \text{thickness} \times (n_{||}-n_{\perp}),$$

where $n_{||}$ and $n_{\perp}$ are the refractive indices of the birefringent structure for light polarized parallel to its long and short axes, respectively. When both components reach the analyzer they

15

are in turn split and only their components that are parallel to the analyzer can pass through. Since the exiting components will be polarized in the same plane they can now interfere. If there was no birefringence, $\Delta\phi$ would be 0 and the components parallel to the analyzer (the ones that can exit the system) would cancel each other out and no light would be detected (this is the same situation as in the absence of specimen, mentioned above). Whenever birefringence makes $\Delta\phi \neq 0$, constructive interference occurs, and the observer can see the birefringent structure. However, even in the presence of birefringence, if the stage is rotated through a complete circle it will pass through four positions at which light disappears because the polarized beam is aligned with one of the object's axes (and there is no beam component aligned with the other axis). Therefore, by rotating the stage, the orientation of the axes (fast and slow) can be determined, but it cannot be decided which is which. To do this, objects of known birefringence, called compensators, are introduced. *Table 3* describes the two most commonly used compensators and their use.

A polarizing microscope usually includes some components relevant to its mode of operation. These are summarized in *Table 4*.

5 Differential interference contrast (DIC) microscopy (Nomarski optics)

This type of microscopy can be viewed as a variation of polarization microscopy and includes both a polarizer and its crossed analyzer. Here, light from the polarizer passes through a Wollaston prism positioned in the condenser's front focal plane, thus generating two parallel beams, each polarized at 45° with respect to the first polarizer. As shown in *Figure 6*, the specimen is therefore illuminated by two coherent beams polarized at 90° to each other, which have a lateral displacement (shear) of approximately the resolution limit of the objective. At this point, the specimen may introduce an optical path difference ($\Delta_{specimen}$) in the region

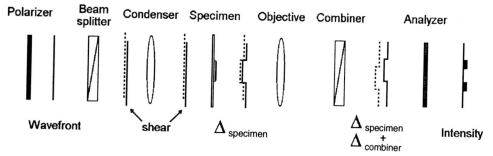

Figure 6. Components and image formation in a differential interference contrast microscope.

of the wavefront that traverses it, affecting both beams. After leaving the objective, the beams are combined again by a second Wollaston prism that removes the shear. This second prism, called the combiner, can introduce a further (adjustable) optical path difference ($\Delta_{combiner}$) between the two beams. As in the polarization microscope, the two orthogonally polarized beams reach the analyzer (with respect to which they are at 45°) and thus the intensity of the exiting beam will depend on the total optical path difference that has been introduced. In the absence of specimen the background intensity will depend on the value of $\Delta_{combiner}$, and background will be black if $\Delta_{combiner} = 0$, or bright (maximum) if $\Delta_{combiner} = \lambda/2$. The combined effect of the two Wollaston prisms is thus equivalent to a birefringent specimen placed in a polarization microscope with its slow and fast axes at $45°$ with respect to the analyzer and introducing a selectable optical path difference ($\Delta_{combiner}$). The presence of the specimen introduces $\Delta_{specimen}$, but affecting *only* the edges (see *Table 5*), resulting in an edge-contrasting effect, the intensity of which depends on the selected $\Delta_{combiner}$. The generation methods for different edge-contrasting effects are summarized in *Table 5*. The magnitude of the shear has to approximate the resolution limit so that only the edges are detected, rather than creating a diffuse halo. Therefore, a different set of Wollaston prisms has to be used for each objective.

6 Interferometer microscope

The interferometer microscope differs from the DIC microscope in that the normal (ordinary) and the reference (extraordinary) beams are separated *by a sizeable distance*. As in the DIC microscope, both beams are generated by a Wollaston prism and finally combined by another Wollaston prism. A $\lambda/2$ plate (for $\lambda = 546$ nm – green light) is placed between the beam splitter and the specimen in order to rotate the beams, so that at the combining step the ordinary beam will become extraordinary and vice versa (see the description of Wollaston prisms in the Glossary). However, one of the beams will form a sharp image while the other will form a severely distorted one where no detail from the specimen can be seen. Provided the specimens are not birefringent, optical path differences between the specimen and the background can be measured accurately by means of a compensator.

7 Hoffman modulation contrast microscopy

This contrast-generating method allows detection of optical gradients within a specimen. The system (*Figure 7a*) features a slit aperture located in the front focal plane of the condenser lens and a modulator in the back focal plane of the objective lens. Therefore, both the slit aperture and the modulator are located in conjugate planes and the former is imaged on the latter. The modulator (*Figure 7b*) has three zones: a dark one (D) with 1% transmittance, a gray one

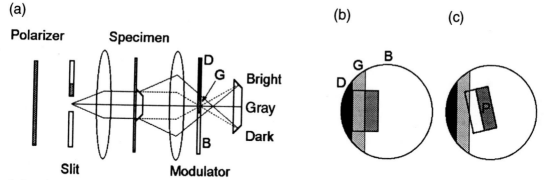

Figure 7. Components and image formation in a Hoffman modulation contrast microscope. (a) Contrast generation; (b) correct and (c) incorrect slit alignment.

Types of Microscopy

(G) with 15% transmittance, and a bright one (B) with 100% transmittance. The slit aperture is focused inside the G zone.

Optical (phase) gradients can be considered as prisms inside the specimen, as depicted in *Figure 7a*. When the specimen is illuminated three situations may arise:

1. Beams encountering no lateral phase gradient in their paths will pass through the G zone of the modulator (where the slit is focused), exiting with 15% of the original intensity.
2. Beams encountering a positive (note that this is arbitrary) lateral phase gradient will be refracted and will pass through the B zone, thus exiting without attenuation.
3. Beams encountering a negative lateral phase gradient will be refracted and will pass through the D zone, thus exiting with 1% of the original intensity. Dashed lines in *Figure 7a* indicate light paths if an optical gradient had not been found.

Therefore, Hoffman modulation contrast produces an image whose intensity is proportional to the first derivative of the phase gradients, and objects appear shadowed on one side, creating a relieved appearance.

The actual system includes two modifications to produce larger resolution and contrast control:

1. Both the slit aperture and the modulator are located off-axis. As a result, illumination is oblique, rather than axial, and resolution is enhanced by approaching the conditions of equation 3b (*Table 1*, Chapter 1).
2. Half of the aperture slit is made of polaroid (P in *Figure 7c* and shadowed in *Figure 7a*) and a polarizer is introduced between the aperture and the light source. The polaroid part of the slit is focused on the B region and thus, by rotating the polarizer, the amount of stray (unmodulated) light in the image can be controlled.

The aspect of the slit aperture and the modulator (as seen through a phase telescope) is shown in *Figure 7b* and *c*. The setting-up procedure for Hoffman modulation contrast is described in *Table 6*. Unstained cells viewed under Hoffman modulation contrast are shown in *Figure 2c* of Chapter 1.

8 Interference reflection microscopy

The interference reflection microscope uses the possibility of generating constructive or destructive interference when light is reflected by surfaces of different refractive indices, as depicted in *Figure 8*. Reflection in medium of higher refractive index, as is usually the case for the specimen in its immersion medium, introduces a $\lambda/2$ phase shift; therefore, if no other phase shifts (Δ) are introduced, the total phase shift ($\Delta\phi$) will be $\lambda/2$ and the resulting beam will destructively interfere with that resultant from reflection in the glass/medium interface. Whenever the distance between specimen and coverslip is large enough, constructive interference will occur and the zone will be viewed as bright. Zones where Δ is close to 0 will thus be seen as dark, as is the case with the contact points between specimen and coverslip.

The system (*Figure 9a*) works in epillumination mode, thus using a half-silvered mirror tilted at 45°, which reflects 50% of the incident light. Note that even if the specimen was a perfect reflector, the collected intensity would be only 25% of the illuminating intensity.

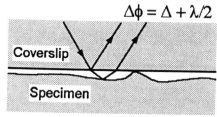

$$\Delta\phi = \Delta + \lambda/2$$

Coverslip

Specimen

Figure 8. Interfering beams in interference reflection microscopy.

9 Fluorescence microscopy

In fluorescence microscopy the specimen is illuminated by light of a certain range of wavelengths (not necessarily

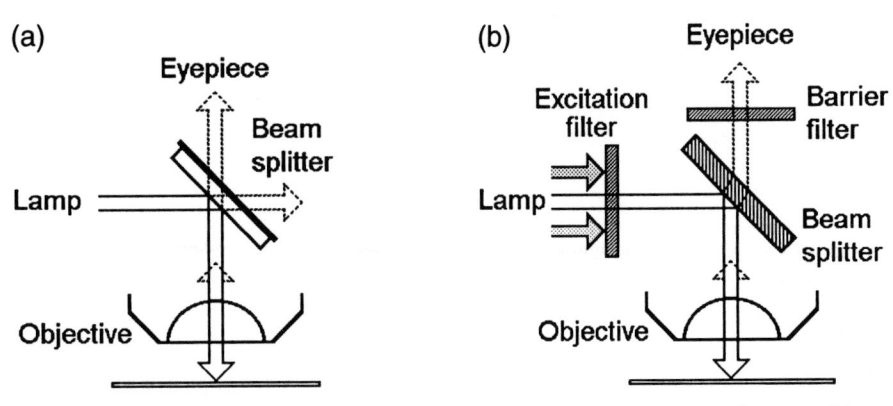

Figure 9. Epillumination assemblies for an interference reflection microscope (a) and a fluorescence microscope (b).

monochromatic) while viewing emitted (fluorescent) light of longer wavelengths. Thus, only fluorescent specimens can be visualized, their fluorescence being either intrinsic or introduced *ad hoc* by means of a suitable labeling procedure. *Figure 9b* depicts the most commonly used arrangement, which uses epillumination. The proper selection of illuminating and emitted wavelengths is performed by means of two filters (barrier and excitation) and a chromatic beam splitter. Since light emitted by labeled specimens has a very low intensity, fluorescence microscopy introduces some special equipment requirements, which are summarized in *Table 7*. Cells viewed under conventional fluorescence microscopy are shown in *Figure 2d* of Chapter 1.

10 Confocal microscopy

Confocal microscopy is intended to achieve a high axial resolution. Contrary to conventional microscopy, it relies on point illumination, rather than field illumination. A schematic diagram of the corresponding configuration is shown in *Figure 10a*. The system usually works in fluorescence mode (it can also be used in reflection mode) and with epillumination. The specimen is illuminated by a point source, consisting of a laser beam focused on a small aperture. Hence, the intensity reaching out-of-focus points on the specimen is lower than using conventional (field) illumination. In turn, fluorescent (or reflected) light leaving the specimen is focused on the small detector aperture. As can be noted in *Figure 10a*, the combined effects of point illumination together with point detection result in a strong rejection of out-of-focus light. The chromatic beam splitter reflects the excitation light towards the specimen, while allowing the emission light to reach the detector. However, since it is a mirror, the optical path from the source aperture to the objective lens is essentially the same as that from the objective lens to the detector aperture. Both the illuminating and detector apertures are focused on the illuminated point in the specimen, meaning that the distance from the illuminated point to the objective and the distance from the source/detector aperture are determined by equation 1 (Chapter 1, *Table 1*). This is the reason why the system is called *confocal*. The detector is a photomultiplier (PMT), because the signal to be detected is usually very low. Also, the light source is most frequently a laser.

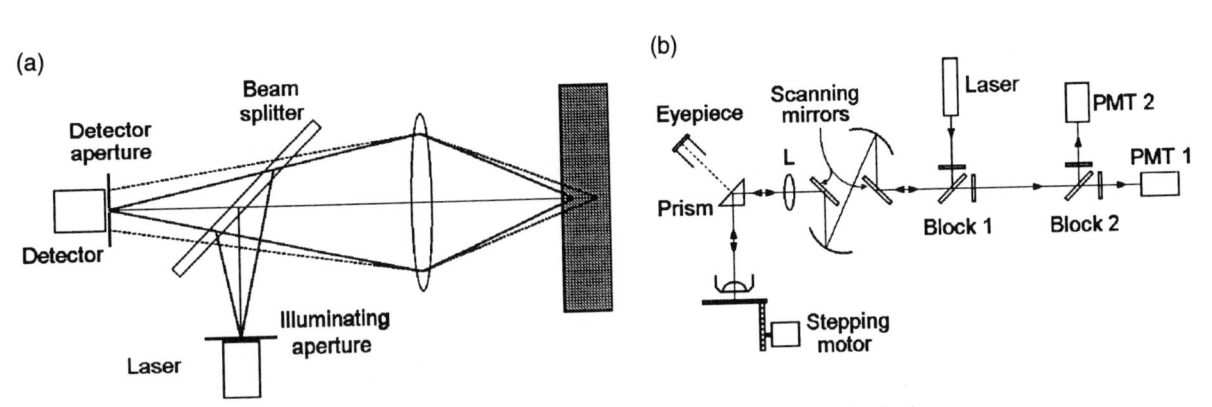

Figure 10. Confocal microscopy. (a) Fundamental assembly; (b) scheme of a laser scanning confocal microscope.

In a confocal system only a single point is viewed at a time. Therefore, in order to build an image, that point has to be scanned over the sample and the measured intensities recorded. This is achieved by either scanning the beam using two mirrors (laser scanning confocal microscope – LSCM) or by moving the stage in the specimen plane (stage scanning confocal microscope – SSCM). LSCM allows fast scanning of biological specimens, whereas SSCM has the advantage of working in optimum focusing conditions (the scanned point is always on the optic axis). Measured intensities are recorded and displayed by a host computer, which also drives the scanning system. All the thin sections along the axis obtained are largely devoid of out-of-focus light, this effect being called *optical sectioning* (see *Figure 2e*, Chapter 1), and repetition of the x–y scanning along the z axis allows the construction of a 3-D image of the object being analyzed. Projection of all z sections over a single plane results in a 2-D image which, unlike conventional microscopy, contains only in-focus information; thus a large depth of field is achieved by adding very thin in-focus sections. This projected image is called *extended focus image* (see *Figure 2f*, Chapter 1). Also, the y mirror can be fixed and scanning performed only with the x mirror and the z motor, thus generating a vertical (z–x section) as in *Figure 2g* of Chapter 1.

The attachment of a LSCM system to a conventional microscope is depicted in *Figure 10b*. The prism allows switching between confocal and normal observation. The laser beam is highly collimated, so is the beam returning to the system (this one by focusing on the focal plane of lens L). Filter block 1 carries a dichroic mirror and emission and barrier filters in order to separate illuminating and collected beams, while block 2 (optional) carries a dichroic mirror and two barrier filters to allow for the introduction of a second PMT. x–y scanning mirrors and the focus knob-attached z stepping motor are also depicted.

Table 1. Role of each component of a compound microscope

Component	Role
Collector lens	Focuses the lamp filament in the front focal plane of the condenser. Likewise, ensures that the filament is not imaged in the specimen plane
Field iris	Limits the size of the illuminated field to the observed region only, thus avoiding scattered light from outer regions of the specimen. Affects neither brightness nor resolution. Conjugate with specimen plane
Condenser iris	Restricts the diameter of the illuminating parallel beam, preventing light from reaching the mounts of the objective lens, thus reducing scattered light. Affects resolution (equation 3a)[a] and depth of field (equation 4)[a] by lowering $NA_{condenser}$ when closed. Conjugate with lamp filament plane
Condenser	Focuses a uniform cone of light on to the specimen. Its aperture can be controlled by the condenser iris, therefore affecting the resolution
Objective	Forms the primary (real) image of the specimen on the back focal plane of the eyepiece. Its NA is the principal determinant of the resolving power of the microscope
Eyepiece	From the primary image, forms an image at infinity which can be observed by a relaxed eye and locates the exit pupil at a comfortable distance

[a]Equations as in Chapter 1, *Table 1*.

Table 2. Procedure for setting Köhler illumination

Step	Comment
1. Center the lamp according to the manufacturer's instructions	—
2. Focus on a well contrasted specimen using a low power objective	The images seen are those of the specimen and the field iris, if slightly closed. The field stop determines the maximum viewable area
3. Focus the image of the field iris by moving the condenser up and down	Both sets of conjugate planes are now established in their corresponding positions
4. Center the image of the field iris by using the centering knobs of the condenser	The illumination is centered
5. Select the working objective	—
6. Open the field iris to just outside the boundaries of the field of view	The whole available field of view is illuminated and no light reaches unviewed zones of the specimen
7. Replace an eyepiece by a phase telescope (or Bertrand lens) and focus it to the back focal plane of the objective	The condenser iris is now viewed, also the lamp filament if the ground glass is removed. Thus, lamp centering can be checked
8. Adjust the condenser iris to about 70% of the back focal plane of the objective	Light is prevented from reaching the mounts of the objective lens
9. If phase contrast is to be used, select and center the appropriate condenser annular aperture (illuminated) with the phase ring (dark) by using the centering knobs	All background (undiffracted) light is made to pass through the $\lambda/4$ ring in the phase plate. Note that in phase contrast there is no condenser iris to adjust. Condenser centering knobs are those that move the whole condenser assembly. Phase contrast centering knobs move only the annular aperture
10. Replace eyepiece and adjust the lamp intensity by varying the lamp voltage	If blue tones are lost, reduce intensity by introducing neutral density filters between the lamp and the condenser. Include blue filter if necessary
11. If the objective has to be changed go to step 5, above	Conjugate planes will remain set and centered. Field and condenser apertures and lamp voltage may need re-adjustment. Annular (phase contrast) aperture may need re-centering

Table 3. Compensators for polarization microscopy

Type	Description	Use
Quartz wedge	Introduces known optical path differences (Δ) by sliding the wedge in/out. If specimen thickness is known, birefringence can be calculated (see text)	With monochromatic light (no specimen), Δ is calibrated (when $\Delta = n\lambda$, background goes black). In white light the specimen will go black if the Δ that it introduces is matched by the wedge's and its axes are crossed with the wedge's. If the specimen does not go black, the axes are parallel and it has to be rotated $90°$
First-order red plate	Introduces a Δ of one wavelength of green light (thus seen as purple-red). A birefringent specimen will add extra Δ and shift the absorption band to other colors	Specimen is rotated and color changes are noted. If slow axes are parallel, color is shifted to blue (absorption shifted to red). If slow axes are crossed, color is shifted to red. A reference color scale can be used to estimate the Δ introduced by the specimen

Table 4. Features and components of the polarization microscope

Component	Function
Strain-free condenser and objectives	Strain-free lenses introduce no birefringence
Rotatable polarizer and analyzer	Allow for system set up
Rotatable stage	Visualization of birefringent structures and determination of their axes' alignment
Compensators	Discrimination of fast/slow axes and measurement of birefringence
Reference cross-wires in the primary image plane	Reference for specimen rotation

Table 5. Image formation in a DIC microscope

$\Delta_{specimen}$	$\Delta_{combiner}$	Total Δ (pattern)	Intensity pattern	Background	Specimen
0	0			Dark	No detail
0	$\lambda/2$			Bright (maximum)	No detail
0	$\lambda/4$			Low intensity	No detail
$\neq 0$	0			Dark	Bright edges. In edges, total $\Delta > 0$
$\neq 0$	$\lambda/2$			Bright (maximum)	Dark edges. In edges, total $\Delta \neq \lambda/2$
$\neq 0$	$\lambda/4$			Low intensity	One edge brighter than background and the other darker. In the bright edge, $\Delta \approx \lambda/2$; in the dark edge, $\Delta \approx 0$

Table 6. Set-up for Hoffman modulation contrast

Step	Comment
Insert a dark blue or green filter into light path	Monochromaticity improves image quality
Set up Köhler illumination	Establishes conjugate planes
Focus phase telescope/Bertrand lens	Allows visualization of slit aperture and modulator
Turn and displace slit aperture until the clear part overlaps with G zone	Undiffracted beams will pass through 15% transmittance zone. *Figure 7b*, properly aligned slit; *Figure 7c*, misaligned slit
Remove phase telescope	—
Close field iris, readjust condenser and open field iris just beyond the field of view	The presence of the slit aperture may require readjustment of condenser focusing
Rotate polarizer until contrast is optimized	Amount of stray light (passing through P and B zones - *Figure 7b* and *c*) is controlled

Table 7. Equipment requirements for fluorescence microscopy

Component	Requirement
Lamp	Must provide a very high luminous density and suitable wavelengths for excitation of the fluorochromes in use, which can be in the UV region. Usually mercury (HBO)
Neutral density filters	Since intensity of mercury lamps cannot be adjusted, it has to be controlled by means of filters inserted in the light path. These are called neutral density filters because they attenuate all wavelengths equally
Filter holders	Different fluorochromes usually require different excitation/emission filter sets. Holders include the excitation and emission filters and the chromatic beam splitters and have to be easy to exchange. Slots for further barrier filters can also be present
Objectives	Must have an NA as large as possible in order to have a high brightness (equation 6[a]). Usually oil-immersion and UV-transparent if UV illumination is to be used. For identical NA, phase contrast objectives will be less bright than bright field ones because of the annular stop. However, phase contrast or other means of contrast rendering optics is highly desirable in order to evidence fluorescence-negative zones (very often too low a contrast for bright field observation)

[a]See Chapter 1, *Table 1.*

Chapter 3 FILTERS AND MIRRORS FOR MICROSCOPY

1 Interference filters and chromatic beam splitters

The most commonly used mirrors and filters are the interference ones. These consist of layers of different materials arranged as in *Figure 1a*. The scheme in *Figure 1a* indicates that both the reflected (solid lines) and transmitted (dashed lines – depicted only for the third layer) beams will interfere with each other. Depending on the refractive indices and thicknesses of the different layers, as well as their numbers, the interference will be destructive for some wavelengths and constructive for others. As a result, very efficient filters can be constructed with a transmittance close to 100% for the passing colors and close to 0 for the blocked ones, while displaying a transmittance vs. wavelength curve with a very high slope at the cut-off wavelength. They can thus be designed to match any application. As opposed to absorption filters (stained glass), the absorption efficiency of interference filters does not depend on filter thickness. Chromatic beam splitters (also called dichroic mirrors) are constructed to reflect certain wavelengths and transmit others. Their reflection/transmission performance is equivalent to that of interference filters.

Clearly, the behavior of both interference filters and chromatic beam splitters will depend on the angle of incidence of the light beam. Filters are to be located normal to the optic axis, while chromatic beam splitters usually will work at 45°.

Types of filters and terms employed are described in *Table 1*.

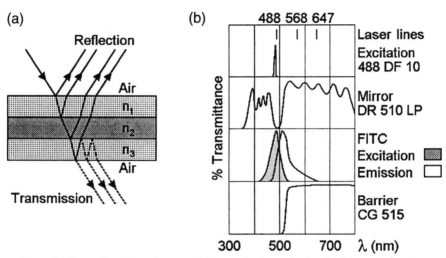

Figure 1. Interference filters. (a) Generation of interference; (b) filter and mirror selection for fluorescein isothiocyanate fluorescence detection with Kr/Ar mixed gas laser epiillumination (BioRad MRC600 laser scanning confocal system).

Filters and Mirrors for Microscopy

2 Choice of filter sets for fluorescence applications

In order to produce high quality images, a filter set has to eliminate any light that does not originate from the desired fluorochrome. The ratio between light intensities for the specific (i.e. fluorochrome-originated) and unspecific (i.e. generated by other components, scattering or reflections) emissions is called signal-to-noise ratio and has to be kept as high as possible. This is achieved by the following procedures:

1. Excitation and emission bands must not overlap.
2. Excitation spectrum must comprise only wavelengths that efficiently excite the desired fluorochrome.
3. Emission spectrum must comprise only wavelengths that include the maximum emission of the desired fluorochrome.
4. Other spectral characteristics such as elimination of ultra-violet (UV) and infra-red (IR) light are usually included.

A wavelength range selection can also be performed by combining two filters or a filter and a dichroic mirror. Tandem-arranged filters have an overall transmittance corresponding to the product of their respective transmittances at each wavelength.

Table 2 describes the filter/dichroic mirror selection for fluorescein visualization under Kr/Ar mixed gas laser excitation in a BioRad MRC600 laser scanning confocal microscope. The corresponding spectra are depicted in *Figure 1b*.

Table 1. Types of interference filters

Filter/mirror	Comment
Excitation filter	Allows entrance into the system of a wavelength range corresponding only to the excitation maximum of the fluorochrome. Prevents excitation of unwanted (auto) fluorescent components
Barrier filter	Allows visualization of the specific wavelength range corresponding to the fluorochrome emission. Prevents detection of unwanted fluorescence sources and scattered/reflected light
Long/short pass filter	Filters passing wavelengths longer (long pass) or shorter (short pass) than a certain specified value. Nomenclature usually specifies the cut-off wavelength and the long/short pass nature, for example, 600LP is a filter passing $\lambda \geq 600$ nm
Band pass filter	Filter passing a range of wavelengths. Nomenclature specifies the center wavelength of the passing range and the half band width (distance in wavelength between the half-maximum transmission points). Other letters may describe manufacturing characteristics. Example: 514 DF 10, a filter passing a 10 nm-wide band centered at 514 nm
Multiband filters	Filters with more than one transmission band, usually employed in simultaneous visualization of two fluorochromes. They are specifically designed to work with a particular fluorochrome combination and dichroic mirror and often for a certain type of light source

Table 2. Selection of filter sets for fluorescein detection according to the spectra of *Figure 1b*

Filter/mirror	Role
Excitation filter 488 DF 10	Filter passing a 10 nm-wide band centered at 488 nm. Prevents other major laser lines (568 and 647 nm) from entering the system
Dichroic reflector DR 510 LP	Mirror transmitting $\lambda \geq 510$ nm (long pass) and reflecting shorter wavelengths. Directs 488 nm laser light towards the specimen and passes fluorescent light (≥ 510 nm) into the system. Note that high transmittance at 350–450 nm is irrelevant
Barrier filter CG 515	Long pass filter transmitting green fluorescence ($\lambda \geq 515$ nm). It does not need to be a band pass filter, because no other emissions at longer wavelengths can be expected for a 488 nm excitation.

Chapter 4 **ABERRATIONS**

Optical systems do not perfectly reproduce the imaged object. Rather, they introduce deviations called aberrations. These can be subdivided in *chromatic* (arising from the fact that the refractive index varies with wavelength) and *monochromatic*. The latter group includes five aberrations: spherical, coma and astigmatism (which blur the image, making it unclear), field curvature and distortion (which deform the image). The aberrations are described in *Table 1* and the corresponding schemes are shown in *Figure 1*.

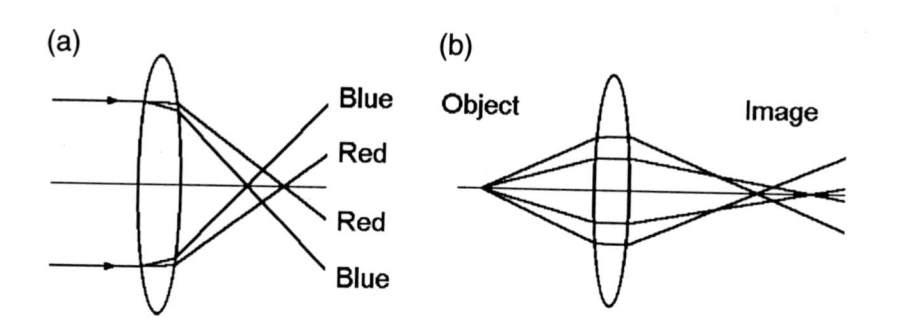

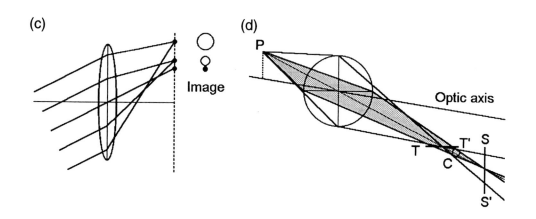

Figure 1. Aberrations. (a) Chromatic; (b) spherical; (c) coma; (d) astigmatism – sagittal plane is shadowed: P, point; S–S′, sagittal focus; T–T′, tangential focus; C, circle of least confusion. Continued overleaf: (e) curvature of field; (f) distortion.

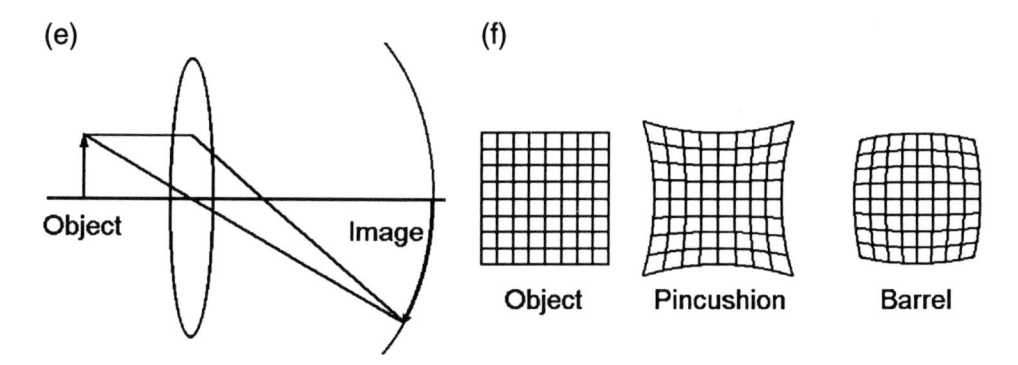

Figure 1. Aberrations. (e) Curvature of field; (f) distortion.

Table 1. Lens aberrations

Name	Cause	Effect	Correction
Chromatic (*Figure 1a*)	Variation of refractive index of glass with wavelength. Rays of different wavelengths are focused on different points	Axial: image surrounded by color fringes Lateral: different magnification for different wavelengths, e.g. red image larger than blue	Combination of lenses of different refractive indices Compensating eyepiece Use of monochromatic light
Spherical (*Figure 1b*)	Rays passing through marginal areas are refracted more than those passing through central ones. No common point of convergence for all rays	No sharp focusing; only 'zone of least confusion'	Combination of positive convex lens with negative concave lens Planoconvex lens with flat side towards surface Aspheric surfaces Some objectives include a correction collar to compensate for different coverglass thicknesses
Coma (*Figure 1c*)	For off-axis points, rays passing through the center of the lens form an image closer to the axis (positive coma) than peripheral ones. These, in turn are not focused on a point but on circles of decreasing intensity	Off-axis points appear as a central dark spot pointing towards the axis (positive coma), accompanied by a comet-like cone of successive wider and less intense circles formed by marginal rays. Magnification depends on distance to the axis	Appropriate design of radii of curvature can eliminate the problem

Continued

Aberrations

Table 1. Lens aberrations, *continued*

Name	Cause	Effect	Correction
Astigmatism (*Figure 1d*)	For off-axis points, rays contained in the meridional plane are focused at a different distance to those in the sagittal plane. The ray connecting the point and the center of the lens and the optical axis define the meridional plane. The sagittal plane is normal to the former	A point is imaged as a circle of least confusion. At shorter distances to the lens, this circle turns into an ellipse and finally a line along the sagittal plane. At longer distances the effect occurs along the meridional plane	Proper combination of concave and convex lenses
Curvature of field (*Figure 1e*)	The image is not formed on a plane but on a spherical surface of radius roughly equal to the focal length	When the center of the image is focused, the periphery becomes blurred, and vice versa	Negative eyepiece (medium-power objectives) Special flat-field, or 'plano', objectives
Distortion (*Figure 1f*)	Magnification varies from the center to the periphery (transverse magnification may be a function of the off-axis distance)	Greater magnification at the center causes 'barrel' distortion. Greater magnification at the periphery causes pincushion distortion	Usually corrected. Should not be noticeable

Chapter 5 LIGHT SOURCES

1 Lamps

Table 1 summarizes the properties of the two basic types of lamps: incandescent and arc lamps. When choosing a microscope lamp, two important features to consider are:

1. *Spectral characteristics.* The filament of an incandescent lamp can be considered as a radiating black body. Thus, its spectrum is described only by specifying the temperature: a rise in filament temperature shifts the emission to shorter (blue) wavelengths. Each 'color curve' is therefore defined by a *color temperature* which is expressed in Kelvins (K). Spectral curves can be found in photomicrography books. Since filament temperature is dependent on applied voltage, this means of controlling light intensity will result in spectral shift (red shift at low voltage). Suppliers often report the corresponding color temperatures for various applied voltages.

In contrast, arc lamps show a band spectrum which may contain a (weaker) continuous background. In some cases (e.g. high-pressure xenon arc lamp – XBO) the spectrum can be assigned a color temperature which can be approximately 6000 K (similar to daylight). The high-pressure mercury arc lamp (HBO) provides very intense spectral lines that match the absorption spectra of certain fluorochromes.

2. *Luminous density.* It is the mean luminous density (rather than the total light output) of the source which determines the illuminance of the field (together with the numerical aperture of the condenser). The highest mean luminous densities are normally achieved with arc light sources which are thus usually employed for applications requiring high field illuminance (e.g. fluorescence microscopy) .

2 Lasers

LASER is the acronym for *L*ight *A*mplification by *S*timulated *E*mission of *R*adiation. In such a system, an excitation source excites a medium, pumping electrons to a high (metastable) energy level. Some of these electrons lose energy, emitting photons that resonate between two mirrors (one of them half-silvered). These photons will in turn trigger the de-excitation of more electrons and thus, the emission of more photons. As a result, highly coherent monochromatic light is emitted through the half-silvered mirror. The main properties of laser light are:

- a high degree of monochromaticity
- a small divergence (parallel beam)
- a high intensity
- a high degree of spatial and temporal coherence
- a polarized emission
- a Gaussian beam profile

The emission bands of some common lasers are described in *Table 2.*

Table 1. Comparison of incandescent and arc light sources

Type	Cause of emission	Spectra	Examples	Uses	Comments
Incandescent	Incandescent filament	Continuous, defined by working temperature (K)	Tungsten–halogen	Bright-field, phase contrast Applications not requiring high intensity	Cheap. Tungsten–halogen show less blackening with age than tungsten lamps
Arc	Ionized gas	Intense bands with some continuous background	High-pressure mercury arc (HBO) High-pressure xenon arc (XBO)	High intensity applications, (e.g. fluorescence – especially HBO lamp)	Expensive. Risk of blow when exceeding life

Table 2. Characteristics of some common lasers

Laser	Bands (nm)
Argon ion	488
	514
He/Ne	543
Kr/Ar mixed gas	488
	568
	647

Chapter 6 **LENS NOMENCLATURE**

The characteristics of an objective lens are usually specified by a three-line legend which, although variable, may look like this

$$\begin{array}{ll} \text{S Plan 20 PL} & (1) \\ 0.46 & (2) \\ 160/0.17 & (3) \end{array}$$

(1) Type of objective (S plan) and magnification (× 20). The former corresponds to a phase contrast objective and thus the type of contrast is also specified: positive low (PL). Some objectives may also specify the phase contrast condenser aperture to use, e.g. Ph 2. Types of objectives usually specify the aberration corrections of the lens and other properties. Examples are:

Apochromatic (Apo): highly compensated for chromatic aberration (three wavelengths) and spherical aberration.

Plan: corrected for field curvature.

UV, Neofluar, Fluo: UV light-transmitting objectives.

LWD–ULWD: long working distance–ultra-long working distance.

PO: polarized light (strain-free).

(2) Numerical aperture (0.46). If it is an immersion objective, the immersion medium is specified in this line (e.g. 1.40 Oil). Multi-immersion objectives will specify this, together with a correction collar for selecting the desired immersion medium.

(3) Tube length in mm (160)/coverslip thickness in mm (0.17). Objectives for variable coverslip thicknesses may specify 0–2, and the proper thickness is set by the user by means of a correction collar. A dash (—) indicates that coverslip thickness is not important.

Some of those characteristics may also be specified in condensers, e.g. numerical aperture, immersion medium (if necessary).

Eyepieces specify the magnification and other characteristics as wide-field, built-in micrometer disc, etc.

Chapter 7 **MEASUREMENT METHODS**

1 Size measurement

Methods for size measurement can be based on three different procedures: comparison with objects of known size; calibration with a graticule of known size (which is alternated with the specimen) and placement of a graticule in the back focal plane of the eyepiece (conjugated with the specimen plane and thus simultaneously focused). The latter method will require calibration with a known standard. The methods are summarized in *Table 1*. Typical sizes of cells and organelles are given in *Table 2*.

2 Counting

Estimation of the number of objects per unit volume is performed by counting them when contained in a precisely defined volume. For this purpose, several chambers have been designed which display a graticule of known dimensions and leave a definite space between its surface and the coverslip (these are usually thick in order to prevent bending by surface tension). Depending on the size of the objects to be counted, chamber sizes range from the Sedgewick Rafter chamber (50×20 mm $\times 1$ mm depth) to the improved Neubauer chamber (hemocytometer) (3×3 mm $\times 0.1$ mm depth, with nine smaller 1×1 mm squares), which is shown in *Figure 1*.

The number of objects counted per area unit (N) follows a Poisson statistical distribution. Hence its standard deviation is $\sqrt{N}$, and therefore the standard error (SE) can be expressed as $SE = N/\sqrt{N}$. *Figure 1* illustrates the counting criterion for cells inside large-graduation 1×1 mm squares;

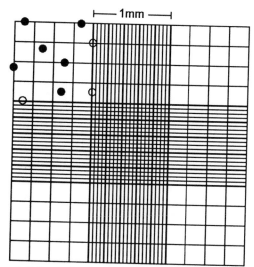

it is common practice to include cells localized on two edges (e.g. upper and left hand – filled circles) and exclude cells localized on the other two edges.

Figure 1. Scheme of an improved Neubauer chamber. Thick lines indicate triple grid. Circles in the upper left 1 × 1 mm square indicate objects to be included (filled) and objects to be excluded (open).

Table 1. Linear measurement methods

Method	Description	Comments
Inclusion of objects of known size in the specimen	Particles can be polystyrene or glass beads, if homogeneous and of known size. In erythrocyte-containing samples these can provide a size reference. Measurements can be performed on a photograph	Accurate
Graticules in the eyepiece	Graticules placed in the back focal plane of the eyepiece are focused on to the image. Grids and circles of various sizes allow size comparison for the viewed objects	Accurate Need calibration
Calibrated slides	Counting chambers like hemocytometers have calibrated grids of known dimensions, allowing direct measurement of object sizes	Rough estimation (the shortest distance between lines is 50 μm)
Fixed dimensions on the microscope	Measuring the size of the field of view can allow a rough estimation of sizes of the contained objects	Very rough
Mechanical stage	Mechanical stages allow movement in both x and y axes. By means of a Vernier scale, displacement can be controlled with an accuracy of 0.1 mm	Accurate to 0.1 mm (Vernier scale) in both x and y axes
Calibrated standards	Stage micrometers (graticules) allow calibration of the system. Since they are not viewed together with the specimen, an image of the graticule must be recorded by means of photograph, TV camera, camera lucida, etc.	Very accurate. Graduations can be as low as 10 μm
Scales in the microscope	*Micrometer eyepieces.* Allow easy exchange of back focal plane graticules and their proper focusing *Filar eyepiece.* Instead of a fixed graticule in the back focal plane, this eyepiece incorporates movable lines that can be accurately displaced by means of a micrometer screw	Very accurate Need calibration Very accurate Need calibration

Table 2. Typical sizes of cells and organelles

Cell or organelle	Dimensions (μm)
Prokaryote cell	0.15–5
Eukaryote cell	10–100
Nucleus	5–25
Mitochondrion	1–10
Chloroplast	2–8
Golgi apparatus	1
Lysosomes and peroxisomes	0.2–0.5
Plant cell wall	0.1–10

Chapter 8 **PHOTOMICROGRAPHY**

The basic equipment for photomicrography includes a 35 mm camera body attached to the microscope by means of a phototube, with an exposure meter. Usually microscopes include special ports for attaching photographic cameras. The following is a general description of typical systems.

Light is diverted from the eyepieces to the camera by means of a prism. The image in the camera must be on the same focal plane (parfocal) as the primary image. This focusing is ensured by the phototube, which is set up during installation and should need no further corrections.

The exposure time is calculated by the exposure meter, according to the film being used. Most exposure meters use *center-weighted average metering*, which means that the meter calculates the average brightness over the whole field of view but biases the value towards that in the center of the field. This exposure time is correct for evenly illuminated fields. However, in cases where samples consist of bright spots on a dark background, as is often the case for fluorescence microscopy, average metering will calculate an inadequate (too long) exposure time. *Spot meters*, which measure brightness in a small area, are thus more appropriate in these cases. If a spot meter is not available, the exposure for these types of pictures has to be adjusted by trial and error, taking pictures at times shorter than that indicated by the average meter.

For film selection, the main criteria are fine grain, medium contrast and sensitivity, which has to be higher the dimmer the specimen is. Film sensitivity is described by the ASA (or ISO) number (also called 'speed' of a film). For low light conditions (e.g. fluorescence microscopy) high ASA numbers (e.g. 800) are used. For color films, the type of light is specified and has to be considered. *Tungsten films* are balanced for the color spectrum of a tungsten lamp (see Chapter 5, Section 1), while *daylight films*, balanced for this spectrum, give the best results with mercury lamps, due to similar spectral characteristics. For example: a daylight film used with tungsten illumination will give a yellowish picture. Thus, it is also important to use the tungsten bulbs at the correct working voltages. Since the spectra of tungsten lamps is rich in red wavelengths, for color photography they can be corrected by means of a blue filter. *Table 1* lists the characteristics of some Kodak films used for microscopy.

Table 1. Characteristics of some Kodak films for use in microscopy

Film	Type	Sensitivity (ASA)	Filter for daylight[a]	Filter for tungsten (3200 K)	Comments
Ektachrome 160T	Tungsten	160	85B	—	Color film. Suitable for bright field microscopy with tungsten lamp
Ektachrome 200	Daylight	200	—	80A	Color film. Suitable for fluorescence microscopy
T-MAX 400	Tungsten/daylight	400	—	—	Black and white film. Very flexible film, for different illumination conditions

[a]Filters refered to are Kodak Wratten gelatin filters for compensating for spectral differences between film and light source.

Photomicrography

Chapter 9 **FIXATION**

The aim of fixation procedures is to preserve a specimen in a state which resembles its natural state as closely as possible. Properties to consider when choosing a fixative are:

1. Size and shape preservation.
2. Penetration. This has to be as fast as possible in order to reach all zones of the specimen simultaneously, minimizing changes during the process.
3. Denaturation. This should be minimal, especially when specific detection methods such as antibodies are to be used.
4. Extraction of components. During fixation, some cellular components may be lost, depending on the method employed. The method to choose is the one that minimizes the elimination of the components to be studied.
5. Autofluorescence. If fluorescence detection methods are to be used, autofluorescence-generating methods (such as glutaraldehyde fixation) have to be avoided.

Fixation methods can be either physical or chemical. *Physical* methods rely on protein denaturation and consist of heat- or air-drying. These are usually applied to smears and can be easily performed. Freezing (liquid N_2, CO_2, etc.) is used for specimen sectioning, usually followed by chemical fixation. *Chemical* methods are based on protein cross-linking or protein denaturation and coagulation. *Table 1* shows examples and properties of some of these fixatives.

Other methods include: picric acid in saturated aqueous or ethanol solutions (e.g. Rossman, Gendre or Bouin), Methacarn (60% methanol, 30% chloroform, 10% glacial acetic acid), periodate–lysine–formaldehyde (0.9% formaldehyde, 75 mM lysine, 10 mM sodium periodate), etc.

Table 1. Properties of some common chemical fixatives

Fixative	Chemical action	Penetration	Effect on cell components	Specimen quality
Ethanol, methanol, acetone, 70–100%	Coagulants. Denature proteins	Fast	Extracts lipids. Destroys organelles	Shrinkage. No autofluorescence
Acetic acid, 5–30%	Protein denaturation	Fast	Destroys organelles. Possible destruction of enzyme activity	Swelling. No autofluorescence
Formaldehyde, 3–4%	Protein cross-linking	Fast, but reaction can be slow	Preserves cellular structures. Possible destruction of enzyme activity. Free CHO groups must be blocked to avoid interaction with subsequent NH_2-containing reagents	Good size preservation. No autofluorescence
Glutaraldehyde, 0.25–4%	Protein cross-linking	Slow	Preserves cellular structures. Effects on enzyme activity. Free CHO groups must be blocked to avoid interaction with subsequent NH_2-containing reagents	Good size preservation. Autofluorescent
Mercuric chloride, 3–6%	Protein coagulation	Fast	Preserves cellular structures. Some effects on enzyme activity	Slight shrinkage. No autofluorescence

Chapter 10 **EMBEDDING AND CUTTING**

Blocks of tissue have to be processed into slices that are thin and transparent enough to be observed under the microscope. Thus, fixation usually has to be followed by dehydration, clearing (these preceded by decalcification in the case of bone specimens), embedding and cutting of the specimen.

1 Dehydration

Dehydration is the first step in the processing of fixed tissues, and has to be performed whenever the solidifying (embedding) medium is water-immiscible. This is usually achieved by sequential use of alcohol aqueous solutions, alcohols, acetone, etc. These solvents either attract water from tissues (hydrophilic) or dilute aqueous tissue fluids. Common dehydrating agents are ethanol, industrial methylated spirit (IMS), methanol and acetone. A general stepwise procedure consists of sequentially immersing the specimen in 50% ethanol, 70% ethanol, 90% ethanol, absolute ethanol, and toluene. Immersion times can vary from 2 to 9 h or overnight, depending on the size of the tissue block.

2 Clearing

For proper observation, specimens must be rendered transparent. This is achieved through a process called *clearing*. Usually, clearing is the last step of the dehydration sequence. The essential requirement for a clearing agent is to be miscible with both the last dehydrating agent and the embedding medium, so that no insoluble (opaque) residues will remain after embedding. Many clearing agents have a refractive index which is similar to that of the components of the specimen (proteins).

In the procedure described in Section 1, the last step (toluene) is the clearing step. Common clearing agents are xylene, toluene, chloroform, paraffin, petrol, 1,1,1-trichloroethane and others.

Note: most clearing agents are highly flammable liquids.

3 Decalcification

For obtaining sections of bone specimens it is necessary to remove the bone minerals and thus soften the tissue. The bone mineral is mainly composed of phosphate, carbonate and calcium, as well as other ions, and therefore mineral removal is achieved by solubilizing calcium (decalcification). Decalcifying agents are acids or calcium-chelating agents. *Table 1* summarizes some decalcifying agents and their properties.

4 Embedding

Tissues have to be embedded in a solid medium that provides the necessary rigidity to enable cutting of very thin sections. After fixation and dehydration, tissue blocks are impregnated by waxes or resinous materials. *Table 2* lists some commonly used embedding media.

The most common embedding medium is paraffin wax. Melting points of paraffin waxes can range from 40 to 70°C and the higher the melting point, the harder the wax will be. For embedding, the wax is heated to 2–3°C above its melting point and poured into an appropriate mold. When the wax starts to solidify over the surface of the mold, the specimen is introduced. When a thin film of semi-solid wax starts to form on the surface, the whole mold is submerged in cold water for fast solidification. Wooden blocks can be attached by melting part of the wax surface with a heated spatula. Also, embedding can be performed under vacuum conditions (400–500 mmHg) in order to

Embedding and Cutting

remove air bubbles in the tissue and rapidly remove the clearing agent.

The whole tissue processing (from fixation to embedding) can be automated by means of processing machines.

5 Cutting

Cutting of specimen sections is performed by means of microtomes. Microtomes employ a very sharp knife to cut sections a few microns thick. The thickness of these sections is adjusted by means of either a screw or a gear system. The different types of microtomes are described in *Table 3*.

Knives can be of different types according to profile (wedge, planoconcave, biconcave or tool edge) or material (high carbon steel and harder materials such as sintered tungsten carbide and diamond). Knife sharpening is a critical process and can be performed using different abrasives, either by hand or by means of automated machines.

Table 1. Summary of decalcifying agents

Type	Examples	Observations
Strong acid	1 M nitric acid 1 M hydrochloric acid	Only for urgent work Have effect on the stainability of the tissue
Weak acid	4 M formic acid Formic acid – formalin	Routine non-urgent samples Decalcification complete in 1–10 days
Chelating agents	Ethylenediaminetetraacetic acid sodium salt (EDTA)	Slow, up to 6–8 weeks Has little or no effect on other specimen components

Table 2. Some commonly used embedding media

Waxes	Resins	Others
Paraffin wax	Acrylic	Agar
Polyethylene glycol	Epoxy	Gelatine
Ester and polyester waxes	Urea–formaldehyde	Celloidin
Microcrystalline wax		

Table 3. Types of microtomes

Type	Operation	Comments
Rocking microtome	Fixed knife. Tissue moving through an arc strikes the knife	Not ideal for hard tissues. Very simple
Rotary microtome	A wheel actuates the forward movement of the block	Suitable for large numbers of sections. Can be motorized. Not for very hard tissues
Base sledge microtome	Block holder on a sledge which slides forward against the knife	Knife angle easily reset. Types for all specimen sizes and degrees of hardness
Freezing microtome	Assembly cooled by CO_2. Knife moving on an immobile tissue block	For frozen-section work. Not useful for paraffin blocks because knife movement prevents formation of ribbons of sections
Vibrating knife microtome	Knife vibrating at 50–60 Hz	Sections may be obtained without fixation, embedding or freezing. Useful for work in non-denaturing conditions

Chapter 11 **STAINING**

The process of staining consists of visual labeling of structures of interest. Its purposes can be to render structures visible; to identify structures by means of a particular color-producing reaction; or analysis of biochemical and/or biophysical properties of certain components. Stains can be fluorescent or non-fluorescent.

1 Non-fluorescent stains

1. *Dyes*. Certain colored compounds can bind to biochemical entities more or less specifically, essentially by non-covalent interactions. For this reason they are often classified as acidic or basic. When the bound dye shows a different absorption spectrum, the phenomenon is called metachromasia. Staining of viable cells with the uptake of stain being a physiologically active process is called vital staining. *Table 1* shows the properties of some common dyes.

2. *Stained products of biochemical reactions*. Some biochemical phenomena (e.g. endogenous enzymes, molecules and metabolic activities) can be used to form a specific colored product. *Table 2* summarizes some of these specific labeling procedures.

3. *Stained products of deliberately introduced enzymes*. This method is essentially the same as the former, but the specificity of the reaction (i.e. the specific localization of the enzymatic activity) is obtained by delivering the enzyme in an antibody-bound form. This methodology is thus called immunohistochemistry, and it allows specific detection of cellular antigens. *Table 3* lists some of the enzyme–substrate combinations that are used for these methods. When two antigens

are detected simultaneously, the procedure is called dual labeling.

2 Fluorescent stains

Fluorochromes are molecules that can absorb light of a certain wavelength to emit light of a longer wavelength. The parameters that characterize a fluorochrome are:

1. The absorption spectrum, which gives information about the wavelengths at which the fluorochrome can be excited. Usually only the absorption maximum is specified;
2. The emission spectrum, which gives information about the wavelength range of its fluorescent light emission. Usually only the emission maximum is specified;
3. The extinction coefficient (ε, cm^{-1} M^{-1}), which relates the amount of absorbed light to the concentration of the fluorochrome in solution;
4. Quantum yield, which indicates the efficiency with which an absorbed photon is converted to emitted light.

The combination of extinction coefficient and quantum yield are the determinant factors of the intensity of a fluorochrome.

For a fluorochrome to be useful in microscopy, its absorption and emission maxima have to be separate enough to be efficiently resolved by the filter/mirror system. This separation is called Stoke's shift. Whenever possible, a fluorochrome must be chosen to have a high extinction coefficient and a high quantum yield, in order to maximize its sensitivity, and a large Stoke's shift, in order to facilitate separation of excitation and fluorescent light. Also, not all of the excitations of electrons will result in the emission of a (fluorescent) photon: some de-excitations can occur by modes that involve a chemical modification of the molecule, resulting in the destruction of its fluorochrome activity. This phenomenon is called photobleaching, because, when extensive, it results in the loss of a fluorescent label.

Fluorescent light emission depends on the de-excitation of electrons and, therefore, its intensity depends only on: (a) the number of excited electrons at a certain time, and (b) the half-life of the excited state (neglecting other non-fluorescent de-excitation modes). The number of excited electrons depends on the intensity of the illuminating (excitation) light, but its maximum value is obviously equal to the total number of fluorochrome molecules. Thus, emission of fluorescent light is a *saturation* phenomenon, indicating that proportionality with the excitation intensity will occur only at low values of the latter.

Subtle changes in spectral properties can be introduced by minimal chemical modification of a basic fluorochrome molecule, such as fluorescein. As a result, 'families' of fluorochromes can be obtained where all of their members display similar absorption/emission spectra that cover different wavelengths. *Figure 1a* depicts the basic structure and numbering of 4,4-difluoro-4-bora-3a,4a-diaza-s-indacene (BODIPY). The various substitutions on the basic BODIPY structure result in a family of fluorochromes,

some of whose members are listed in *Table 4*. *Figure 1b* shows the spectral variations for some BODIPY derivatives.

3 Fluorescent analogs of biomolecules

3.1 Reactive groups for the synthesis of fluorescent analogs

Fluorochromes can be covalently linked to macromolecules, resulting in a molecule that maintains both the spectral properties of the fluorochrome and the (main) biochemical properties of the original macromolecule. Such a compound is called a *fluorescent analog* of the macromolecule. To obtain such a binding, basic fluorochrome molecules are substituted with reactive groups that will react with certain groups present in biomolecules. For example, the fluorochrome fluorescein is often employed in the form of fluorescein isothiocyanate (FITC), which covalently binds to amine groups. Thus, the choice of a particular derivative of a fluorochrome depends on the group in the macromolecules that it is desired to

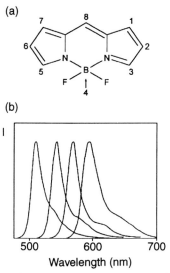

Figure 1. Spectral variations within the BODIPY family of fluorochromes. (a) Basic structure and numbering of BODIPY; (b) some spectra observed within the BODIPY family. Reprinted with permission from Molecular Probes, Inc.

react with; usually a very abundant group, in order to achieve high labeling ratios (number of fluorochrome molecules per macromolecule). *Table 5* summarizes the reactive groups that are most commonly used for covalently linking fluorochrome molecules to biomolecules.

3.2 Fluorescent probes coupled to biomolecules

The fluorochrome with which to prepare a fluorescent analog is usually chosen in terms of its spectral characteristics, according to the properties discussed in Section 2. *Table 6* lists fluorochromes that are most commonly used in a biomolecule-bound form.

4 Fluorescent physiological indicators and tracers

4.1 Ion and pH indicators

Some fluorochrome molecules can change their spectral characteristics when placed at different pH or when bound to certain ions (e.g. Ca^{2+}). As an example, SNARF-1 has an emission maximum at 585 nm in its acid form and an emission maximum at 635 nm in its basic form. The ratio between 585 nm vs. 635 nm emissions depends *only* on the pH of the medium, which can thus be measured if both emissions can be quantitated. Therefore, no pH reference has to be included in the sample. The same is also valid for certain ion-dependent probes, which are thus called *ratiometric*. Again, these probes do not require the presence of an ion concentration reference solution, and ion concentration can be measured by ratioing the intensities of the high ion and low ion emissions. *Figure 2* shows spectra of Indo-1 at increasing Ca^{2+} concentrations. *Table 7* lists some ratiometric probes.

Some probes show fluorescence only when bound to certain ions. Thus, no ratiometric analysis can be performed. However, these *non-ratiometric* probes can still be used for ion measurements, provided it is possible to include a reference (non-ion-dependent) fluorochrome *which has the same distribution inside the specimen* as the ion-dependent

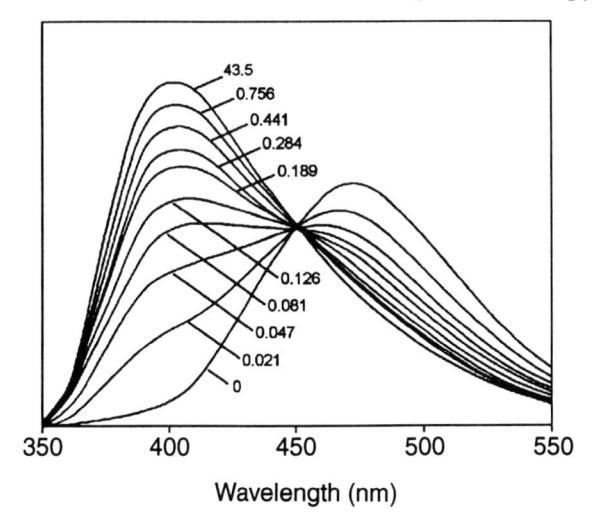

Figure 2. Spectra of Indo-1 at increasing concentrations of Ca^{2+} (μm). Excitation: 355 nm. Reprinted with permission from Molecular Probes, Inc.

probe. For quantitation with these probes, the presence of a known concentration of the measured ion is necessary.

4.2 Probes for viability assays

Fluorochrome molecules that can discriminate between live and dead cells usually rely on cell membrane potential, which is maintained only in live cells. The reason for this is that many electrically neutral probes can passively cross the cell membrane while charged probes cannot. *Table 8* lists some of such fluorescent viability tests.

4.3 Site-specific probes

Site-specific probes can selectively label a certain subcellular compartment either by themselves or by coupling to another molecule (not necessarily an antibody). While labeling of cytoskeleton (actin) is frequently achieved by fluorescently labeled phalloidin, labeling of organelles frequently relies on fluorescently labeled lipid molecules. DNA- and RNA-labeling probes usually intercalate in the nucleic acid strands. *Table 9* lists some organelle-specific probes.

4.4 Photoactivatable ('caged') compounds

Caged probes are compounds that do not display a certain physicochemical property until they undergo photolysis. Most frequently, these compounds are non-fluorescent and are rendered fluorescent by photolysis. However, other biomolecules can be 'caged', such as cyclic nucleotides, etc. These compounds are useful in that they allow the generation of a certain signal (e.g. fluorescence, release of bioactive molecules, etc.) when and where desired, by localized flash photolysis. The production of caged probes basically consists of adding a probe of a 'caging group' such as 4,5-dimethoxy-2-nitrobenzyl. Caging groups can usually be cleaved by photolysis at 340–360 nm.

Table 1. Properties of some common dyes

Group	Dye[a]	Type	Use
Triaryl methanes	Basic fuchsin (545)	Basic	Schiff's reagent. Nuclear
	Pararosaniline (545)	Basic	Nuclear
	Crystal violet (589–593)	Basic	Nuclear, vital and Gram stain
	Acid fuchsin (540–546)	Acid	Cytoplasmic
	Aniline blue (600)	Amphoteric	Cytoplasmic
	Fast green (625)	Amphoteric	Cytoplasmic
Hematein	Hematein (445)	Amphoteric	Nuclear hematoxylin dye component
Anthraquinones	Alcian blue (380)	Basic	Mucopolysaccharides
	Alizarine blue (–)	Acid	Nuclear mordant dye metachromatic stain
Xanthenes	Pyronin Y (552)	Basic	RNA/DNA
	Acridine orange (467–497 fluorescent)	Basic	Nucleoli
	Eosin Y (515)	Acid	Cytoplasmic counterstain
Azines	Neutral red (530)	Basic	Vital stain
Oxazines	Nile blue (640)	Basic	Fat, sulfate, differentiates neutral fats and fatty acids
	Cresyl violet (600)	Basic	Metachromatic vital stains
	Brilliant cresyl blue (630)	Basic	Vital stain. Acid mucopolysaccharides

Thiazines	Azure A (630)	Basic	Nuclear
	Azure B (650)	Basic	Nuclear
	Methylene blue (660)	Basic	Nuclear and vital stain
	Toluidine blue (620)	Basic	Nuclear
Monoazo-	Janus green B (400, 610–623)	Basic	Vital stain
Diazo-	Sudan black B (600)	Basic	Fat
	Amido black (blue)	Acid	Nuclear and protein stain
	Oil red O (550)	Acid	Fat
Triazo-	Naphthol yellow S (420)	Acid	Protein
	Dinitrofluorobenzene (400)	Acid	Protein

[a]Numbers in parentheses indicate absorption maxima.

Table 2. Examples of biochemical phenomena used for specific staining

Process/component	Reaction	Comments
Periodic acid–Schiff (PAS)	Periodic acid oxidation of sugars, followed by reaction with reduced basic fuchsin (Schiff's reagent)	Demonstration of polysaccharides
Feulgen reaction for DNA	HCl hydrolysis followed by Schiff's reagent	Purines are eliminated by hydrolysis and aldehydes are reacted by Schiff's reagent
Succinate dehydrogenase	Reduction of nitroblue tetrazolium in the presence of succinate	Reduction yields a blue formazan deposit. Methyl green can be used as nuclear counterstain
Acid phosphatase	α-Naphthyl phosphate in the presence of fast garnet GBC at pH 5.0	The enzyme produces α-naphthol which, coupled to fast garnet GBC, yields an insoluble colored product that reveals the presence of the enzyme. Demonstrates lysosomes

Table 3. Some enzyme–substrate combinations for immuno-histochemistry

Enzyme	Substrate
Peroxidase	Diaminobenzidine–H_2O_2 4-Chloro-1-naphthol–H_2O_2 Tetramethyl benzidine–H_2O_2
Alkaline phosphatase	Naphthol AS BI–fast red TR Naphthol AS BI–fast blue BB Naphthol AS TR–new fuchsin–$NaNO_2$
Glucose oxidase	Nitroblue tetrazolium-β-D-glucose
Galactosidase	Potassium ferricyanide–potassium ferrocyanide–5-bromo-4-chloro-3-indolyl-β-D-galactose

Table 4. Substituents and resulting spectra for various BODIPY fluorophores

Fluorophore[a]	Substituents[b]
BODIPY 493/503	1-Alkyl, 3-alkyl, 5-alkyl, 7-alkyl, 8-alkyl
BODIPY 503/512	3-Alkyl, 5-alkyl, 7-alkyl
BODIPY 542/563	3-Alkyl, 5-(p-methoxyphenyl)
BODIPY 564/570	3-Alkyl, 5-styryl
BODIPY 581/591	3-Alkyl, 5-(4-phenyl-1,3-butadienyl)

[a]Numbers correspond to absorption/emission maxima.
[b]Numbers according to *Figure 1a*

Table 5. Some reactive groups for the synthesis of fluorescent analogs of biomolecules

Group to be reacted	Reactive group	Example
-SH	Haloacetyl derivatives Maleimides	Fluorescein iodoacetamide Tetramethyl rhodamine-5-maleimide
-NH_2	Isothiocyanates Succinimidyl esters Sulfonyl chlorides	Fluorescein isothiocyanate (FITC) 5-Carboxyfluorescein succinimidyl ester Texas red sulfonyl chloride

Table 6. Some commonly used fluorophores

Fluorochrome	Absorption max. (nm)	Emission max. (nm)
Fluorescein-5-isothiocyanate (FITC isomer I)	494	520
Tetramethyl rhodamine isothiocyanate (TRITC)	541	572
Rhodamine X isothiocyanate (XRITC)	578	604
Texas red (sulfonyl chloride derivative of sulforhodamine)	596	620
Lissamine rhodamine B	567	584
Eosin-5-isothiocyanate	524	548
Phycoerythrin-R	480–565	578
Lucifer yellow CH lithium salt	428	540
Cascade blue	375, 398	424
Hoechst 33342	340	450
4',6-Diamidino-2-phenylindole hydrochloride (DAPI)	350	470
Ethidium bromide	510	595
Propidium iodide	536	623
Acridine orange	480 (+DNA)	520
	440–470 (+RNA)	650
Pyronin Y	549–561 (+dsDNA)	567–574
	560–562 (+dsRNA)	565–574
	497 (+ssRNA)	563
Rhodamine 123	556	577

Table 7. Examples of ratiometric probes

Probe	Ion	Low ion[a]		High ion[a]	
		Absorption (nm)	Emission (nm)	Absorption (nm)	Emission (nm)
Fura-2, pentapotassium salt	Ca^{2+}	362	512	335	505
Fura red, tetraammonium salt	Ca^{2+}	472	645	436	640
Indo-1, pentapotassium salt	Ca^{2+}	346	475	330	408
Carboxy SNARF-1	pH	518, 548 (low)	587	579 (high)	635
2',7'-bis-(2-carboxyethyl)-5-carboxyfluorescein (BCECF)	pH	457, 482 (low)	520	508 (high)	531

[a]For pH probes, indicates low and high pH

Table 8. Fluorescent viability tests

Probe	Mode of action
Fluorescein diacetate (FDA)	Non-fluorescent and non-charged. Permeates cell membrane and internal esterases (live cells) cleave it into fluorescein, which is fluorescent and membrane impermeant
Acridine orange	Permeates cell membrane. Green/orange fluorescence when bound to DNA/RNA. Live cells show green fluorescence
Ethidium bromide	Membrane impermeant for live cells. Binds to DNA of dead cells showing fluorescence. Used in combination with acridine orange, live cells show green fluorescence and dead cells red fluorescence

Table 9. Examples of organelle-specific probes

Organelle	Probe[a]
Endoplasmic reticulum	$DiOC_6$ (3)
	$DiOC_5$ (3)
Golgi	BODIPY ceramide
	NBD ceramide
	BODIPY sphingomyelin
	NBD sphingomyelin
Mitochondria	Rhodamine 123
	Rhodamine 6G
	Tetramethyl rhodamine, ethyl ester
	Nonyl acridine orange
	DASPEI
	2-Di-1-ASP (DASPMI)
	4-Di-1-ASP (DASPMI)
	EDCK
Nerve terminals/ synaptosomes	4-Di-2-ASP
	RH 414
	FM 1–43
Plant mitochondria	$DiOC_7$ (3)

[a] See Abbreviations list.

Chapter 12 SPECIFIC DELIVERY OF STAINS: IMMUNOHISTO-CHEMISTRY AND IMMUNOFLUORESCENCE

1 General procedures

For detailed analysis of microscopic preparations it is often necessary to visualize components that are not differentially labeled by any stain or fluorochrome. In those cases, immunoglobulins (or their fractions) can specifically deliver different types of labels to a desired cellular location. Common antibody-mediated labels are enzymes and fluorochromes. Also immunogold is a commonly used method, where the label consists of small but detectable antibody-bound gold particles. Labels can be directly attached to the antibody that reacts against the antigen of interest. However, it is common practice to use an unlabeled first (cellular antigen-specific) antibody and a labeled second antibody that specifically reacts with the first one. By this means, a larger amount of label can be attached to each antigen molecule (amplification) and, also, greater flexibility is achieved, due to the possibility of using the same labeled second antibody for a variety of first antibodies.

Table 1 summarizes some of the factors that should be taken into account when choosing an antibody for immunodetection.

The protocol given in *Table 2* for immunofluorescent labeling of surface antigens of human lymphoid cells includes comments on the significance of each step. It should be taken as a general guide, since adaptation for each particular case will most probably be necessary.

When sections are mounted on slides, the procedure is similar to that in *Table 2*, but antibodies are added in an appropriate volume to cover the sample and incubations are carried out in a humid chamber, to prevent evaporation.

Specimen preparation for immunotechniques often requires permeabilization in order to render internal structures accessible to the immunoglobulins. This can be achieved by the fixation method itself if it employs organic solvents (methanol, acetone), or by addition of detergents (Triton X-100, Nonidet P40) which will remove membrane components. Also, saponin permeabilization has been used to preserve cell membranes while allowing internalization of antibodies.

2 (Strept)avidin–biotin methods

Antibody labeling is often accompanied by the use of avidin–biotin (or streptavidin–biotin) systems. Avidin contains four high-affinity (10^{15} M^{-1}) binding sites for biotin; arranged in two pairs. Each pair of avidin sites can only bind one biotin molecule at a time, therefore acting as bivalent. However, since biotinylation usually introduces several biotin groups per protein molecule, the use of biotin–avidin complexes allows amplification of each original antigen–antibody binding event, resulting in a stronger signal. Fluorochrome- or enzyme-labeled avidin is thus a general reagent which can be used to amplify reactions involving biotinylated antibodies. For the formation of avidin–biotin complexes the same protocol is used as for antigen–antibody interactions.

3 Enzyme–anti-enzyme methods

An increasingly used methodology that enhances the sensitivity of immunohistochemical techniques is the enzyme–anti-enzyme strategy, of which the most frequent combinations are peroxidase–anti-peroxidase (PAP) (exemplified here) and alkaline phosphatase–anti-alkaline phosphatase (APAAP). This strategy relies on the amplification that can be achieved by sequential addition of antibody molecules, which are bivalent. The anti-peroxidase complex consists of anti-peroxidase antibody (most frequently a monoclonal antibody) bound to horseradish peroxidase in a previous incubation. The antigen in the specimen is reacted with an antibody (first antibody) as described in Section 1, and a *bridging antibody* (polyclonal) is included which can bind both the first antibody and the anti-peroxidase antibody (this by being species-specific) and is responsible for the amplification. The sequence of reactions is as described in *Table 3*.

The former enzyme–anti-enzyme strategy can be subjected to modifications by inclusion of a biotin–(strept)avidin system. For instance, if both the enzyme and the bridging antibody are biotinylated, the enzyme can be linked by adding avidin between steps 3 and 4 (*Table 3*); and in this case the technique is called the ABC method (avidin–biotinylated peroxidase complex). Considering that biotinylation reactions introduce several biotin molecules per protein molecule, the method also introduces amplification. However, the advantage of the avidin–biotin link is that the bridging antibody is no longer required to react with *both* the first and the anti-enzyme antibodies, thus rendering the system more flexible.

Table 1. Properties of immunoglobulins and immunoglobulin fractions for use in immunohistology

Reagent	Advantages	Disadvantages
Polyclonal immunoglobulins	Less denaturation-sensitive due to the large number of epitopes recognized	Possible high cross-reactivity and non-specific binding
Monoclonal immunoglobulins	High specificity. Usually low non-specific binding	If the recognized epitope is altered by specimen processing, the antibody cannot be used
Immunoglobulin fractions, F(ab')$_2$	Low non-specific binding. No reaction with Fc receptors	Expensive

Table 2. Protocol for immunofluorescent labeling of surface antigens

Procedure	Comments
1. Suspend cells in PBS with added 5% inactivated human AB serum (PBS + S)	Serum proteins will reduce the non-specific binding of antibodies. Inactivated AB serum is used in order to avoid possible reactivity against cells
2. Add first antibody at the appropriate dilution in PBS + S. Incubate for 45 min in cold (4°C)	Incubation in cold prevents internalization of antigen–antibody complexes
3. Wash twice with PBS + S	—
4. Incubate second antibody under the same conditions as the first one	—
5. Wash twice with PBS + S	—
6. Wash once with PBS only	Removes protein to allow efficient formaldehyde fixation
7. Add 4% formaldehyde in PBS and incubate for 5 min	Fixation renders cells resistant to mounting
8. Wash once with PBS. Using a cytocentrifuge, load cells on to a glass slide. Mount with an appropriate agent. Add coverslip and seal edges with nail varnish	Mounting medium prevents dehydration and photobleaching

Table 3. The peroxidase–anti-peroxidase (PAP) method

Step	Reaction	Comments
1. Formation of PAP complex	$Ab_p + HRP \rightarrow Ab_p{:}HRP$	HRP is added in excess so that no free Ab_p remains. If this happens, non-HRP-bound Ab_p would bind to Ab_b, thus diminishing the sensitivity. Since each Ab_p can bind two HRP molecules, another level of amplification is introduced here
2. Reaction of Ab_1 with Ag (specimen)	$\|Ag + Ab_1 \rightarrow \|Ag{:}Ab_1$	Ag in the specimen is detected by Ab_1. At this point the *specificity* of the reaction is determined
3. Introduction of Ab_b	$\|Ag{:}Ab_1 + Ab_b \rightarrow \|Ag{:}Ab_1{:}Ab_b$	Many Ab_b react with each Ab_1 molecule, thus producing the *amplification* effect Ab_b is added in excess, such that Ab_1 is saturated. If this does not happen, the number of Ab_b binding sites available for binding the PAP complex will be less than optimum, thus losing amplification efficiency
4. Reaction with PAP complex	$\|Ag{:}Ab_1{:}Ab_b + Ab_p{:}HRP$ $\rightarrow \|Ag{:}Ab_1{:}Ab_b{:}Ab_p{:}HRP$	The whole complex is formed and enzymatic color reaction can be performed

Ag, antigen; Ab_1, first (anti-antigen) antibody; Ab_b, bridging antibody (polyclonal); HRP, horseradish peroxidase; Ab_p, anti-HRP antibody.

Notes:
Steps 2 and 3 are followed by washings, as described in *Table 2*.
Ab_b has to be able to recognize *both* Ab_1 and Ab_p.
Note that step 1 is incubated separately from steps 2 and 3. The PAP complex is added in step 4.
Since, as with Ab_1, several Ab_b molecules can be bound to Ab_p, after step 4 further amplification can be achieved by repeating step 3 with the addition of more Ab_b, which will now bind to Ab_p. Although the sensitivity will thus be higher, the reaction may show an increase in non-specific binding.

Chapter 13 DNA PROBES FOR *IN SITU* HYBRIDIZATION

Single-stranded DNA will bind a complementary strand of DNA or RNA. Thus, it is possible to use such a reagent to analyze for the presence of certain DNA (e.g. viral DNA or particular genomic regions) or RNA molecules (e.g. detection of mRNA). In order to evidence the reaction, the single-stranded DNA probe usually contains biotinylated nucleotides which will subsequently bind fluorochrome- or enzyme-labeled avidin. Biotinylated DNA probes are usually produced by nick translation in the presence of biotin-11-deoxyuridine triphosphate.

Chapter 14 ANTI-FADING AGENTS AND SPECIMEN MOUNTING

As mentioned in Chapter 11, Section 2, fluorochromes can undergo photobleaching. This process is dependent on the presence of molecular oxygen interacting with excited fluorochromes. Therefore, some antioxidants can act as anti-fading reagents by consuming molecular oxygen. The most commonly used anti-fading agents are 1,4-diazabicyclo [2.2.2] octane (DABCO), *p*-phenylenediamine, *N*-propylgallate and hydroquinone.

Mounting agents also include glycerol, which prevents dehydration. A drop of mounting agent is layered over the specimen and the coverslip applied and sealed with nail varnish.

A mounting agent can be prepared containing 50% glycerol, 0.1% NaN_3 and 100 mg ml^{-1} DABCO in PBS.

Chapter 15 SUPPLIERS OF EQUIPMENT AND CHEMICALS

1 Microscopes and ancillary equipment

Bio-Rad Laboratories,
Bio-Rad House, Maylands Avenue, Hemel Hempstead, Herts HP2 7TD, UK
Tel. (0442) 232552
Fax (0442) 259118

Cambridge Instruments,
PO Box 123, Buffalo, NY 14240-0123, USA
Tel. (716) 8913000
Fax (716) 8913080

Carl Zeiss Scientific Instruments,
Zeiss England House, 1 Elstree Way, Borehamwood, Herts, WD6 1NH, UK
Tel. (081) 9531688
Fax (081) 9539456

Goodfellow Metals Ltd,
Cambridge Science Park, Milton Road, Cambridge, CB4 4DJ, UK
Tel. (0223) 69671
Fax (0223) 316975

Graticules Limited,
Morley Road, Tonbridge, Kent TN9 1RN, UK
Tel. (0732) 359061

Modulation Optics Inc.,
100 Forest Drive at Esat Hills, Greenvale, NY 11548, USA
Tel. (516) 4848882
Fax (516) 4845921

UK Representative: Micro Instruments Ltd, Long Hanborough, Oxon OX7 2LH. Tel. (0993) 883595; fax (0993) 883616

Nikon UK Ltd,
Haybrook, Halesfield 9, Telford, Salop TF7 4EW, UK
Tel. (0952) 587444
Fax (0952) 588009

Nikon Inc.,
623 Steward Avenue, Garden City, NY 11530, USA
Tel. (516) 2220200
Fax (516) 2220267

Olympus Optical Co. (UK) Ltd,
2–8 Honduras Street, London EC1 Y0TX, UK
Tel. (071) 2532772
Fax (071) 2516330

Olympus Corporation,
4 Nevada Drive, Lake Success, NY 11042-1179, USA

Omega Optical, Inc.,
PO Box 573, 3 Grove Street, Brattleboro, Vermont 05302, USA
Tel. (802) 2542690
Fax (802) 2543937

Wild Leitz UK Ltd,
Davy Avenue, Knowlhill, Milton Keynes, MK5 8LB, UK
Tel. (0908) 666663
Fax (0908) 609992/3

2 Chemicals, stains and antibodies

Accurate Chemical and Scientific Corp.,
300 Shames Drive, Westbury, NY 11590, USA
Tel. (800) 6456264
Fax (516) 9974948

Astell Scientific,
172 Brownhill Road, London SE6 2DL, UK
Tel. (081) 3004311
Fax (081) 3002247

BDH Laboratories, Merck Ltd,
Merck House, Poole, Dorset, BH15 1TD, UK
Tel. (0202) 669700
Fax (0202) 665599

Biomeda,
PO Box 8045, Foster City, CA 94404-8045, USA
Tel. (800) 3418787
Fax (415) 3412299

UK Distributor: Biogenesis Ltd, 12 Yeomans Park, Yeomans Way, Bournemouth BH8 0BJ. Tel. (0202) 522895; fax (0202) 530367

Calbiochem Novabiochem (UK) Ltd,
3 Heathcoat Building, Highfields Science Park, University Boulevard, Nottingham NG7 2QJ, UK
Tel. (0602) 430840
Fax (0602) 430951

Dako Ltd,
16 Manor Courtyard, Hughenden Avenue, High Wycombe, Bucks HP13 5RE, UK
Tel. (0494) 452016
Fax (0494) 441553

Janssen Pharmaceutical,
Hyde Park House, Cartwright Street, Newton, Hyde,
Cheshire, SK14 4EH, UK
Tel. (0613) 679277
Fax (0613) 678165

Molecular Probes Inc.,
PO Box 22010, Eugene, OR 97402-0414, USA
Tel. (503) 4658300
Fax (503) 3446504

UK Representative: Cambridge BioScience, 25 Signet
Court, Swann's Road, Cambridge CB5 8LA. Tel. (0223)
316855; fax (0223) 60732

Pierce Chemical Co.,
3747 North Meridian Road, PO Box 117, Rockford, IL
61105 USA
Tel. (800) 8-PIERCE or (815) 9680747
Fax (800) 8425007

Pierce & Warriner (UK) Ltd,
44 Upper Northgate Street, Chester, Cheshire CH1 4EF, UK
Tel. (0800) 252185
Fax (0244) 373212

Serotec,
22 Bankside, Station Approach, Kidlington, Oxford OX5
1JE, UK
Tel. (0865) 379941
Fax (0865) 373899

Shandon Southern Instruments,
515 Broad Street, Drawer 43, Sewickley, PA 15143-0043, USA
Tel. (412) 7418400

UK Distributor: Life Sciences International (Europe) Ltd,
Chadwick Road, Astmoor, Runcorn, Cheshire, WA7 1PR.
Tel. (0928) 566611; fax (0928) 565845

Sigma Chemical Co.,
PO Box 14508, St Louis, MO 63178, USA
Tel. (800) 8487791

UK Office: Fancy Road, Poole, Dorset, BH17 7BR. Tel.
(0202) 733114; fax (0202) 715460.

Vector Laboratories, Ltd,
16 Wulfric Square, Bretton, Peterborough PE3 8RF, UK
Tel. (0733) 265530
Fax (0733) 263048

FURTHER READING

1 General texts

Books containing technical data are in bold type.

Bancroft, J.D. and Stevens, A. (1977) **Theory and Practice of Histological Techniques,** 2nd Edn. Churchill Livingstone, London.

Bayliss High, O. (1984) *Lipid Histochemistry*, Royal Microscopical Society Microscopy Handbooks no. 6. Oxford University Press, Oxford.

Bradbury, S. (1991) *Basic Measurement Techniques for Light Microscopy*. Oxford University Press, Oxford.

Dealtry, G.B. and Rickwood, D. (1992) **Cell Biology Labfax**. BIOS Scientific Publishers, Oxford.

Haken, H. (1985) *Light, Vol. 2. Laser Light Dynamics*. North-Holland Physics Publishing, Amsterdam.

Haughland, R.P. (1992–94) **Handbook of Fluorescent Probes and Research Chemicals.** Molecular Probes, Inc., Eugene, OR.

Hecht, E. and Zajact, A. (1987) *Optics*, 2nd Edn. Addison-Wesley, Reading, MA.

James, J. and Tas, J. (1984) *Histochemical Protein Staining Methods,* Royal Microscopical Society Microscopy Handbooks no. 4. Oxford University Press, Oxford.

Lacey, A.J. (1989) **Light Microscopy in Biology**, 1st Edn. IRL Press, Oxford.

Nomenclature Committee of the RMS (1989) *RMS Dictionary of Light Microscopy,* Royal Microscopical Society Microscopy Handbooks no. 15. Oxford University Press, Oxford.

Pawley, B. (ed.) (1990) **Handbook of Biological Confocal Microscopy**. Plenum Press, New York.

Ploem, J.S. and Tanke, H.J. (1987) *Introduction to Fluorescence Microscopy,* Royal Microscopical Society Microscopy Handbooks no. 10. Oxford University Press, Oxford.

Polak, J.M. and Van Noorden, S. (1987) *An Introduction to Immunocytochemistry: Current Techniques and Problems,* Royal Microscopical Society Microscopy Handbooks no. 11. Oxford University Press, Oxford.

Rawlins, D.J. (1992) *Light Microscopy.* BIOS Scientific Publishers, Oxford.

Robinson, P.C. and Bradbury, S. (1992) ***Qualitative Polarized-Light Microscopy,*** Royal Microscopical Society Microscopy Handbooks no. 9. Oxford University Press, Oxford.

Spencer, M. (1982) *Fundamentals of Light Microscopy.* Cambridge University Press, Cambridge.

Thomson, D.J. and Bradbury, S. (1987) *An Introduction to Photomicrography,* Royal Microscopical Society Microscopy Handbooks no. 13. Oxford University Press, Oxford.

Tijssen, P. (1985) ***Practice and Theory of Enzyme Immunoassays***. Elsevier Science Publishers BV, Amsterdam.

Van Noorden, C.J.F. and Frederiks, W.M. (1992) ***Enzyme Histochemistry,*** Royal Microscopical Society Microscopy Handbooks no. 26. Oxford University Press, Oxford.

2 Articles and manuals

Sigma ImmuNotes (various issues). Sigma Chemical Co., St Louis, MO.

Bio-Rad (1991) *Bio-Rad MRC-600 Laser Scanning Confocal Imaging System* (operating manual). Bio-Rad Laboratories, Hemel Hempstead, Herts, UK.

GLOSSARY

Aperture iris: an element limiting the illuminated area of a lens. Limits resolution and brightness of an image. Located in focal planes.

Bertrand lens: a lens inserted in a microscope's optical path which, in combination with the eyepiece, forms a phase telescope.

Birefringence: anisotropy causing the existence of two different refractive indices along two corresponding crystalline axes. Thus, polarized light may travel at two different speeds, depending on whether its polarization plane is parallel or perpendicular to the crystal's optic axis. With $n_{\parallel}$ and $n_{\perp}$ being the respective refractive indices, birefringence is measured as b.r. $= n_{\parallel} - n_{\perp}$.

Conjugate planes: equivalent planes in an optical system, such that an object located in one of them will be focused in all subsequent ones.

Depth of field: the distance through the object along the optical axis within which object's features appear acceptably sharp.

Dichroism: selective absorption of one of the two orthogonal plane-polarized components of an incident beam.

Epillumination: an illumination mode where the illuminating light reaches the specimen from the objective lens, which thus works both as condenser and objective.

Field iris: an element limiting the size of the object that can be imaged. Does not affect brightness and/or resolution. Located in planes conjugate with the object plane.

Focal plane: the plane normal to the optic axis of a lens, that contains a focus (foci are defined by equation 2, Chapter 1, *Table 1*). The front focal plane is that on the object side, while the back focal plane is that on the image side.

Iris (iris diaphragm): a diaphragm of variable size inserted in the optical path.

Numerical aperture (NA): lens parameter defined as n sin θ, where n is the refractive index of the immersion medium and θ is the maximum half angle subtended from an axial point on the focused plane to the lens.

Optic axis: the axis of central symmetry of an optic system.

Optical tube length: the distance from the back focal plane of the objective to the front focal plane of the eyepiece. Standardized to 160 mm.

Paraxial: the region close to the optical axis, where the approximation sin θ = θ is valid (called paraxial, first-order or Gaussian approximation).

Parfocal: objectives with very similar or identical focusing distances, so that very little or no focus adjustment has to be made when changing them.

Phase telescope: a telescope that replaces an eyepiece for inspection of the objective's back focal plane.

Polaroid: the commercial name of a variety of dichroic plastic sheets used for generating polarized light.

Pupil: an image of the aperture iris. It is called entrance pupil when viewed from an axial point on the object side and exit pupil when viewed from an axial point on the image/observer side. For visualization, the pupil of the eye of the observer is located at the center of the exit pupil.

Refractive index: ratio of the speed of light in vacuum to that in matter. Varies with wavelength. The most commonly used relative refractive index is the ratio of the former (absolute) refractive indices of a material and air. Governs the angle of refraction by Snell's law.

Wollaston prism: a beam splitter made of a double prism of birefringent crystal. The incident beam is split into two beams which exit the prism cross-polarized with respect to each other and at different angles.

Working distance: the distance from the objective to the coverslip.

APPENDIX

Care and maintenance of the light microscope

- A good understanding of the optical systems is convenient for optimum results.
- When not in use, replace the microscope in its box, or cover with the appropriate plastic cover.
- The microscope stand must be kept clear. Dust removed with clean cloth. Corrosive liquids removed with cloth moistened in water. Immersion oil, grease, etc., removed with xylene.
- Principal contaminants of optical glasses are: dust; water and aqueous solutions from specimens; immersion oil; and finger-marks. Dust should be blown by dry air from a rubber bulb (not breathing). Aqueous solutions are removed by gentle cleaning with lens tissue moistened in distilled water. Immersion oil and finger-marks should be removed with xylene, not with alcohols since these may dissolve the lens cement (xylene should also be used in low amounts).
- Always remove immersion oil immediately after use, since it hardens upon drying.
- Clean lenses only externally. Never disassemble an objective.
- Care should be taken not to scratch lens surfaces. For focusing, bring the objective as close to the specimen as possible by visual control of the distance. Then, looking through the eyepiece, focus while moving the objective away from the specimen.
- Mercury lamps:
  Never use them beyond their useful life, since they may explode.
  If a mercury lamp explodes, evacuate the room immediately and close the door.

Always keep the room ventilated when the lamp is on, due to ozone production.

Never look directly at a burning mercury lamp.

As a general rule, when the lamp is turned on, do not turn off within 30 min. Conversely, when turned off, do not turn on again within 30 min.

Ensure that lamp ventilation complies with the manufacturer's instructions.

INDEX

ABC method, 75
Aberration, 37–42, 46
  chromatic, 37, 38, 41, 46
  monochromatic, 37
  spherical, 37, 38, 41, 46
Absorption
  filter, 32
  maximum, 61, 69–71
  spectrum, 60, 61
Acetic acid, 54, 55
Acetone, 55, 74
Acid dye, 66
Acid fuchsin, 66
Acid phosphatase, 68
Acridine orange, 66, 70, 71
Acrylic resin, 59
Actin, 65

Agar, 59
Air-drying, 54
Alcian blue, 66
Aldehydes, 68
Alizarine blue, 66
Alkaline phosphatase, 69
Amido black, 67
Amine group, 59, 62, 69
Amphoteric dyes, 66
Amplification, 73, 75, 76
Analyzer, 15–18, 28
Ancillary equipment, 81
Angle of incidence, 32
Aniline blue, 66
Annular aperture, 10, 11, 14, 27
  stop, 31
Anthraquinones, 66

Anti-fading, 80
Antibody, 60, 65, 73–78, 83
Antigen, 3, 60, 73–78
Antioxidant, 80
APAAP, 74
Aperture, 4
  annular, 10, 14, 27
  condenser, 7, 13
  confocal, 23
  detector, 23, 24
  illuminating, 23, 24
  numerical, 5, 6, 26, 31, 43,
    46, 47
  slit, 19, 20, 30
Apo, 46
Apochromatic, 46
Arc lamp, 43, 45

Argon ion laser, 45
Articles, 87
ASA number, 52
Aspheric surface, 41
Astigmatism, 37, 39, 42
Autofluorescence, 2, 35, 54, 55
Avidin, 74, 75, 79
Axial resolution, 5, 23
Axis
  fast, 15, 16, 18, 28
  optical, 1, 28
  slow, 15, 16, 18, 28
Azines, 66
Azure, 66

Back focal plane, 1, 7, 9, 10, 14, 19,
    26, 27, 48, 50
Background
  intensity (DIC), 18, 29

light, 10, 27
Band pass filter, 35
Barrel distortion, 40, 42
Barrier filter, 22, 23, 25, 31,
    34–36
Base sledge microtome, 59
Basic dye, 66
Basic fuchsin, 66, 68
BCECF, 71
Beads
  glass, 50
  polystyrene, 50
Beam
  coherent, 16
  collimated (confocal), 25
  extraordinary, 18
  Gaussian profile, 44
  interfering, 21
  normal, 18

ordinary, 18
parallel, 6, 10, 16, 26, 44
reference, 18
reflected, 32
transmitted, 32
Beam splitter, 12, 17, 18, 22–24,
    31, 32
Bertrand lens, 10, 12, 27, 30
Biomolecule, 62, 63, 69
Biotin, 74, 75
Biotin-11-deoxyuridine triphosphate,
    79
Biotinylation, 74
Birefringence, 2, 15, 16, 18, 28
2′,7′-bis-(2-carboxyethyl)-5-
    carboxyfluorescein, 71
Black body, 43
Block, filter, 24, 25
Blue filter, 27

BODIPY, 62, 63, 69
  ceramide, 72
  sphingomyelin, 72
Bone, 56
Bouin, 54
5-Br-4Cl-3-indolyl-β-D-galactose, 69
Bridging antibody, 75, 76
Bright field, 2, 7
  cells observed in, 3
  objectives in fluorescence microscopy, 31
Brightness, 26, 31
  image, 5, 6
Brilliant cresyl blue, 66

Ca^{2+}, 63, 71
Caged compounds, 65
Caging group, 65

Calibrated
  slide, 50
  standard, 50
Calibration, 48
Camera, 12, 52
  lucida, 50
5-Carboxyfluorescein succinimidyl ester, 69
Carboxy SNARF-1, 71
Cascade blue, 70
Cell, 48, 49, 51
  dead, 65
  lymphoid, 73
  membrane, 65, 71, 74
  viable, 60, 65
Celloidin, 59
Center-weighted average metering, 52
Centering
  illumination, 27

  knobs, 13, 27
Chamber, 48, 50
Chelating agent, 58
Chemical fixation, 54
Chemicals, 81, 83
Chloroform, 54
Chloroplast, 51
CHO groups, 55
Chromatic aberration, 37, 38, 41, 46
  axial, 41
  lateral, 41
Chromatic beam splitter, 23, 31, 32
Circle of least confusion, 39, 42
Clearing, 56
4-Cl-1-naphthol, 69
CO_2, 54
Coagulant, 55
Coagulation, 54
Coherence, 44

Coherent
  beams, 16
  illumination, 5
  light, 44
Collector, 7, 8, 9, 26
Collimated beam (confocal), 25
Color
  film, 53
  reaction, 76
  spectrum, 53
  temperature, 43
Coma, 37, 39, 41
Combiner, 17
Compensator, 16, 18, 28
Compound microscope, 2, 7, 8, 26
Computer (confocal microscopy), 25
Concave lens, 41, 42
Condenser, 7, 8, 9, 13, 15, 26, 27, 28,
  30, 47

  aperture, 7, 13, 46
  dark field, 11, 15
  iris, 6, 9, 26, 27
Confocal
  cells observed in, 3
  image, 25
  microscope,
    laser scanning, 24, 25
    stage scanning, 25
  microscopy, 2, 23–25
  resolution, 5
Conjugate plane, 6, 7, 8, 10, 19, 26,
  27, 30
  illuminating, 10
  image-forming, 10
Constructive interference, 10, 16,
  21, 32
Contact point, 21
Conventional microscopy, 23

Convergence, 41
Convex lens, 41, 42
Correction collar, 41, 46
Corrosive liquids, 90
Counting, 48
  chamber, 50
Coverglass thickness, 41, 46
Coverslip, 21, 48, 80
Cresyl violet, 66
Cross-reactivity, 76
Cross-wire, 28
Crystal violet, 66
Curvature of field, 40, 42
Cut-off wavelength, 32, 35
Cutting, 56
Cytocentrifuge, 77
Cytoplasmic counterstain, 66
Cytoplasmic stain, 66
Cytoskeleton, 65

DABCO, 80
DAPI, 70
Dark field, 2, 11
    illumination, 11
DASPEI, 72
DASPMI, 72
Daylight, 43
    film, 52
De-excitation, 61
Dead cell, 65, 71
Decalcification, 56, 58
Dehydration, 56, 80
Depth of field, 5, 6, 26
    confocal, 25
Destructive interference, 10, 21, 32
Detection, 60
Detector, 24
    aperture, 23, 24

Detergent, 74
4',6-Diamidino-2-phenylindole hydrochloride, 70
Diaminobenzidine, 69
Diaphragm, field, 7
2-Di-1-ASP, 72
4-Di-1-ASP, 72
4-Di-2-ASP, 72
1,4-Diazabicyclo[2.2.2]octane, 80
Diazo-, 67
DIC – *see* Differential interference contrast
Dichroic
    mirror, 25, 32, 35
        selection, 33, 34
    reflector, 36
Differential interference contrast, 13, 16, 17, 29
Diffracted light, 11, 15

Diffracting structures (dark field), 10
Diffraction, 10
4,5-Dimethoxy-2-nitrobenzyl, 65
Dinitrofluorobenzene, 67
DiOC5, -6, -7, 72
Distance, focal, 1, 4, 5, 6, 8
Distance to lens, 1, 5
Distortion, 37, 40, 42
Divergence, 44
DNA, 65, 68, 70, 71, 79
    probes, 79
    stain, 66
Drying, 54
dsDNA, 70
dsRNA, 70
Dual labeling, 61
Dust, 90
Dyes, 60

EDCK, 72
Edge
  -contrasting, 18
  in DIC, 29
EDTA, 58
Electrons, 44
Embedding, 56
Emission maximum, 69–71
Emission band, 33
Endoplasmic reticulum, 72
Energy levels, 44
Enzyme, 68, 69
  activity, 55
  –anti-enzyme, 74
Eosin Y, 66
Eosin-5-isothiocyanate, 70
Epillumination, 2, 21–23, 34
Epitopes, 76
Epoxy resin, 59

Equipment, 81
Erythrocyte, 50
Ester waxes, 59
Esterases, 71
Ethanol, 55
Ethidium bromide, 70, 71
Ethylenediaminetetraacetic acid,
  58
Eukaryote, 51
Evaporation, 74
Excitation, 31
  band, 33
  filter, 22, 23, 25, 31, 34–36
  spectrum, 33
Excited state, 62
Exit pupil, 7
Exposure
  meter, 52
  time, 52

Extended focus, 3
  image, 25
Extinction coefficient, 61
Extraordinary beam, 18
Eye, 7, 26
  pupil, 9, 26
Eyepiece, 6, 7, 8, 9, 12, 26, 27, 41, 47,
  48, 50, 52, 90
  filar, 50
  micrometer, 50

F(ab')$_2$, 76
Fast axis, 15, 16, 18, 28
Fast blue BB, 69
Fast garnet GBC, 68
Fast green, 66
Fast red TR, 69
Fat stain, 66
Fatty acids, 66

Fc receptors, 76
FDA, 71
Ferricyanide, 69
Ferrocyanide, 69
Feulgen reaction, 68
Field
  curvature, 37, 40, 42, 46
  depth of, 5, 6, 26
    confocal, 25
  diaphragm, 7
  illuminated, 26
  illumination, 23
  iris, 6, 8, 13, 14, 26, 27, 30
  of view, 6, 7, 27, 30, 50, 52
  stop, 8, 27
Filament, 7, 26, 27, 43, 45
Filar eyepiece, 50
Film, 52
  black and white, 53
  daylight, 53
  grain, 52
  sensitivity, 52
  speed, 52
  tungsten, 53
Filter, 12, 32, 33, 36
  absorption, 32
  band pass, 35
  barrier, 22, 23, 25, 31, 34–36
  block, 24, 25
  blue, 27
  excitation, 22, 23, 25, 31, 34–36
  holder, 31
  in Hoffmann modulation contrast, 30
  interference, 32, 34, 35
  long pass, 35, 36
  multiband, 35
  neutral density, 27, 31
  selection, 33, 34, 36
  sets, 33, 36
  short pass, 35
  tandem, 33
Finger-marks, 90
First-order red plate, 28
FITC – *see* Fluorescein isothiocyanate
Fixation, 54, 56
  chemical, 54
Fixative, 54, 55
Flash photolysis, 65
Flat-field objective, 42
Fluo, 46
Fluorescein, 33, 36, 62, 71
  diacetate, 71
  iodoacetamide, 69
  isothiocyanate, 34, 62, 69, 70
Fluorescence, 33, 45, 54, 71
  cells observed in, 3
  microscopy, 2, 21–23, 31, 43, 52

Fluorescent
  analog, 62, 63, 69
  light, 21, 36
  stain, 61
Fluorochrome, 31, 33, 35, 61–63, 70, 73, 80
Fluorophore, 69, 70
FM 1-43, 72
Focal
  distance, 1, 4, 5, 6, 8
  length, 42
  plane, 1, 6, 25
    back, 1, 7, 9, 10, 14, 19, 26, 27, 48, 50
    front, 1, 7, 10, 19, 26
Focus, 1, 4
  extended, 3, 25
  knob, 13, 25
  sagittal, 39
  tangential, 39
Focusing, 90
Formaldehyde, 54, 55, 77
Formalin, 58
Formazan, 68
Formic acid, 58
Freezing, 54
  microtome, 59
Front focal plane, 1, 7, 10, 19, 26
Fura red, 71
Fura-2, 71

Galactosidase, 69
Gaussian beam profile, 44
Gelatine, 59
Gendre, 54
Glass
  beads, 50
  ground, 27
Glossary, 88
Glucose oxidase, 69
Glutaraldehyde, 54, 55
Glycerol, 80
Gold particles, 73
Golgi apparatus, 51, 72
Gradient
  optical, 18, 20
  phase, 20
Gram's stain, 66
Graticule, 48, 50
Grease, 90
Grid, 50
Ground glass, 27

Half band width, 35
Half-life, 62
Half-silvered mirror, 21, 44
Haloacetyl derivatives, 69

HBO lamp, 31, 45
He/Ne laser, 45
Heat drying, 54
Hematein, 66
Hematoxylin, 66
Hemocytometer, 48, 50
High-pressure mercury arc lamp, 43, 45
High-pressure xenon arc lamp, 43, 45
Hoechst 33342, 70
Hoffman modulation contrast, 18–20, 30
    cells observed in, 3
Horseradish peroxidase, 75, 76
Human AB serum, 77
Humid chamber, 74
Hybridization, 79
Hydrochloric acid, 58
Hydrogen peroxide, 69
Hydrolysis, 68
Hydroquinone, 80

Illuminance, 43
Illuminated field, 26
Illuminating
    aperture, 23
    rays, 7, 8
Illumination, 5, 7
    centering, 27
    coherent, 5
    dark field, 11
    field, 23
    incoherent, 5
    Köhler, 7, 8, 10, 11, 27
    laser, 5
    point, 23
Image, 4, 5
    at infinity, 6, 7, 26
    brightness, 5, 6
    confocal, 25
    3-D, 25

extended focus, 3
formation, 1, 6
    dark field, 15
    DIC, 17, 29
    Hoffman modulation contrast, 19
    phase contrast, 14
    polarization, 15
-forming rays, 7, 8
illumination, 5
plane, 4, 6, 10
primary, 6, 7, 26, 28, 52
real, 6, 7, 26
size, 6
Immersion, oil, 6, 90
    condenser, 11
    medium, 4, 6, 21, 46, 47
    objective, 6, 31, 46
Immunodetection, 73
Immunofluorescence, 73

Immunoglobulin, 73, 74, 76
Immunogold, 73
Immunohistochemistry, 60, 69, 73
Incandescent lamp, 43, 45
Incoherent illumination, 5
Indo-1, 64, 71
Infinity, image at, 6, 7, 26
Infra-red, 33
Intensity
  confocal microscopy, 23, 25
  DIC, 17, 18, 29
  Hoffman modulation contrast, 20
  interference reflection microscopy,
    21
  laser, 44
Interface, 21
Interference, 11, 34
  constructive, 10, 16, 21, 32
  destructive, 10, 21, 32

filter, 32, 34, 35
  reflection microscopy, 21, 22
Interfering beams, 21
Interferometer microscope, 18
Internalization, 77
Ionized gas, 45
Iris
  condenser, 6, 9, 26, 27
  field, 6, 8, 14, 26, 27, 30
ISO number, 52
Isothiocyanates, 69

Janus green B, 67

Kelvin, 43
Knife, 58, 59
Knob
  centering, 27
  focus, 25

Köhler illumination, 7, 8, 10, 11, 27, 29
Kr/Ar mixed gas laser, 33, 34, 45

Labeling, 60
  dual, 61
  ratio, 63
Lamp, 9, 26, 30, 43
  arc, 43
  filament, 7, 26, 27
  HBO, 31
  incandescent, 43
  intensity, 27
  mercury, 31, 53
  voltage, 27
Laser, 23–25, 33, 34, 36, 44, 45
  beam, 25
  illumination, 5
  scanning confocal microscope, 24,
    25, 33, 34

101

Lens
  cement, 90
  concave, 41, 42
  convex, 41, 42
  nomenclature, 46
  planoconvex, 41
  thin
    image formation, 1, 6
    properties, 4
Lenses in compound microscope, 7
Light
  -absorbing mask, 11, 14
  background, 10, 27
  coherent, 44
  diffracted, 11
  intensity, 43
  microscope, 13
  monochromatic, 28, 41, 44
  path, 1
  polarized, 15, 46
  scattered, 26
  source, 24, 43
  stray (Hoffman modulation
    contrast), 20
Linear measurement, 50
Lipids, 55
Liquid N_2, 54
Lissamine rhodamine B, 70
Live cells, 71
Long
  pass filter, 35, 36
  working distance, 46
Lucifer yellow CH, 70
Luminance, 5
Luminous density, 31, 43
LWD – *see* Long working distance
Lysine, 54
Lysosomes, 51, 68

Magnification, 5–7, 41, 42, 46, 47
  total, 7
Maintenance, 90
Maleimides, 69
Measurement, 48, 50
  linear, 50
Membrane potential, 65
Mercuric chloride, 55
Mercury, high-pressure arc lamp, 43, 45
Mercury lamp, 31, 53, 90
  explosion, 90
  ventilation, 90
Meridional plane, 42
Metachromasia, 60
Metachromatic, 66
  stain, 66
Metastable, 44
Methacarn, 54
Methanol, 54, 55, 74

Methyl green, 68
Methylene blue, 66
Microcrystalline wax, 59
Micrometer, 47
  eyepiece, 50
  screw, 50
Microscope
  compound, 7, 26
  confocal, 23–25
  conventional, 23, 25
  interference reflection, 21
  interferometer, 18
  light, 13
  maintenance of, 90
  type selection, 1
Microtome, 58, 59
Mirror, 32, 34, 36, 44
  dichroic, 25, 32, 35
  half-silvered, 21, 44

scanning, 24, 25
  selection, 34
Mitochondria, 51, 72
Modulator, 19, 20, 30
Molecular oxygen, 80
Monoazo-, 67
Monochromatic
  aberrations, 37
  light, 28, 41, 44
Monochromaticity, 30, 44
Monoclonal immunoglobulins, 76
Mordant dye, 66
Motor, stepping, 24, 25
Mounting, 77, 80
Mucopolysaccharide stain, 66
Multi-immersion objective, 46
Multiband filter, 35

Nail varnish, 77, 80

α-Naphthol, 68
Naphthol AS BI, 69
Naphthol AS TR, 69
Naphthol yellow S, 67
α-Naphthyl phosphate, 68
NBD ceramide, 72
NBD sphingomyelin, 72
Neofluar, 46
Nerve terminals, 72
Neubauer chamber, 48, 49
Neutral density filter, 27, 31
Neutral red, 66
New fuchsin, 69
Nick translation, 79
Nile blue, 66
Nitric acid, 58
Nitroblue tetrazolium, 68
Nitroblue tetrazolium-β-D-glucose, 69
Nomarski, 2, 16

Non-specific binding, 76
Nonidet P-40, 74
Nonyl acridine orange, 72
Normal beam, 18
Nuclear counterstain, 68
Nuclear dye, 66
Nucleoli stain, 66
Nucleotides, 65, 79
Nucleus, 51
Numerical aperture, 6, 26, 43, 46, 47
    condenser, 5
    objective, 5, 31

Object, 1, 4–6, 50
    plane, 6
Objective, 6–9, 13–15, 25, 27, 28,
        31, 41, 42, 46, 90
    immersion, 46
    multi-immersion, 46

Observer, 7
Oil immersion, 6, 31
    condenser (dark field), 11
Oil red O, 67
Optical
    axis, 1, 25
    gradients, 18, 20
    path (confocal), 23
    path difference, 16–18, 28
    section, 3
    sectioning, 25
Ordinary beam, 18
Organelle, 48, 51, 55, 65, 72
Out-of-focus light, 23, 25
Oxazines, 66
Ozone, 90

PAP, 74
    complex, 76

Paraffin wax, 59
Parallel
    beam, 6, 10, 16, 44
    rays, 7
Pararosanilin, 66
Paraxial, 1
Parfocal, 52
PAS, 68
PBS, 77, 88
Penetration, 54, 55
Periodate, 54
Periodic acid, 68
    –Schiff (PAS), 68
Permeabilization, 74
Peroxidase, 69, 74, 75, 76
    –anti-peroxidase, 74, 76
Peroxisomes, 51
pH, 63, 71
Phalloidin, 65

Phase
  advance, 11
  contrast, 2, 10, 11, 27, 31
    condenser, 11
    microscope, 14
    objective, 11, 46
    positive low, 46
  difference, 10, 15
  gradient, 20
  plate, 11, 14, 27
  ring, 10, 13, 27
  shift, 21
  telescope, 10, 20, 27, 30
Phase contrast
  cells observed in, 3
  negative, 11, 14
  positive, 11, 14
p-Phenylenediamine, 80
Photoactivatable compounds, 65

Photobleaching, 61, 80
Photograph, 50
Photolysis, 65
Photomicrography, 52
Photomultiplier, 24
Photon, 44, 61
Phototube, 12, 52
Phycoerythrin-R, 70
Physiological indicator, 63
Picric acid, 54
Pincushion distortion, 40, 42
Plan, 46
Plane
  conjugate, 6, 7, 8, 10, 19, 26, 27, 30
  back focal, 1, 7, 9, 10, 14, 19, 26, 27, 48, 50
  front focal, 1, 7, 10, 19, 26
  image, 4, 6
  object, 6

  plane, 42
  sagittal, 39, 42
  specimen, 7, 26, 48
Plano objective, 42
Planoconvex lens, 41
Plant cell, 51
  mitochondria, 72
Plate, $\lambda/2$, 18
Point
  illumination, 23
  source, 23
Poisson distribution, 48
Polarization microscopy, 11, 28
Polarized light, 2, 15, 44, 46
Polarizer, 11, 15, 16, 17, 19, 28, 30
Polaroid, 11, 20
Polyclonal immunoglobulins, 76
Polyester waxes, 59
Polyethylene glycol, 59

Polysaccharides, 68
Polystyrene beads, 50
Primary image, 6, 7, 26, 28, 52
Prism, 52
  Wollaston, 16, 17, 18
Probe, 63, 64, 71, 72
  DNA, 65
  ion-dependent, 64
  lipid, 65
  RNA, 65
  site-specific, 65
Prokaryote, 51
Propidium iodide, 70
N-propylgallate, 80
Protein
  coagulation, 55
  cross-linking, 54
  denaturation, 54
  stain, 66

Pupil
  exit, 7, 9, 26
  eye, 9
Purines, 68
Pyronin Y, 66, 70

Quantum yield, 61
Quartz wedge, 28

Ratiometric, 64
Ray-tracing, 1
Rayleigh's criterion, 5
Rays
  illuminating, 7, 8
  image-forming, 7, 8
  parallel, 7
Reactive group, 69
Real image, 6, 7, 26
Reduction, 68

Reference beam, 18
Reflected light, 2, 35
Reflection, 21, 23, 32–34
Reflection microscopy
  interference, 21
Refractive index, 5, 11, 21, 32, 37, 41
Resin, 59
Resolution, 5, 6, 26
  axial, 5, 23
  in a confocal microscope, 5, 23
  lateral, 5
  limit, 5, 16, 18
Resolving power, 26
Retardation, 10
Retina, 7, 8
RH 414, 72
Rhodamine 123, 70, 72
Rhodamine 6G, 72
Rhodamine X isothiocyanate, 70

RNA, 70, 71, 79
  stain, 66
Rocking microtome, 59
Rossman, 54
Rotary microtome, 59

Sagittal
  focus, 39
  plane, 39, 42
Saponin, 74
Staining, 60
Saturation, 62
Scale, 50
Scanning, 25
  confocal microscope
    laser, 24, 25
    stage, 25
  mirror, 24, 25
Scattered light, 26, 35

Scattering structures (dark field), 10
Schiff's reagent, 66, 68
Section, 74
  optical, 3
  vertical, 3, 25
Sectioning
  optical, 25
Sedgewick–Rafter chamber, 48
Sensitivity, 61, 74
  enzyme–anti-enzyme, 76
Shear, 16–18
Short pass filter, 35
Shrinkage, 55
Signal-to-noise ratio, 33
Single-stranded DNA, 79
Site-specific probe, 65
Size
  measurement, 48
  preservation, 54, 55

Slide, calibrated, 50
Slit aperture, 19, 20, 30
Slow axis, 15, 16, 18, 28
SNARF-1, 63
Sodium azide, 80
Sodium nitrite, 69
Spatial coherence, 44
Specificity, 76
Specimen, 7, 8, 10, 21, 27, 48, 50, 54, 56
  image, 6, 8
  labeled, 23
  plane, 7, 25, 26, 48
  processing, 76
  quality, 55
  rotation, 28
Spectrum, 43, 52, 69
  emission, 33
  excitation, 33
Spherical aberration, 37, 41, 46

Spot meter, 52
ssDNA, 79
ssRNA, 70
Stage, 13, 16
  mechanical, 50
  micrometer, 50
  rotatable, 28
  scanning confocal microscope, 25
Staining, 68
Standard
  calibrated, 50
  deviation, 48
  error, 48
Stepping motor, 24, 25
Stoke's shift, 61
Stop, 6
  annular, 31
  field, 8, 27
Strain-free, 28, 46

Stray light (Hoffman modulation
    contrast), 20, 30
Streptavidin, 74, 75
Substituents, 69
Substrate, 69
Succinate, 68
  dehydrogenase, 68
Succinimidyl ester, 69
Sudan black B, 67
Sugars, 68
Sulfonyl chlorides, 69
Sulforhodamine, 70
Suppliers, 81
Surface antigen, 3, 77
Swelling, 55
Synaptosomes, 72
Synthesis, 69

Tandem filters, 33

Tangential focus, 39
Telescope, phase, 10, 20, 27, 30
Temperature, 43
Temporal coherence, 44
Tetramethyl benzidine, 69
Tetramethyl rhodamine, 72
  isothiocyanate, 70
  -5-maleimide, 69
Texas red, 70
  sulfonyl chloride, 69
Texts, 86
Thiazines, 66
Thickness (and birefringence), 15,
    28
Thin lens – *see* Lens, thin
Tissue, 56, 58
Toluidine blue, 67
Tracer, 63
Transmission, 34

Transmittance, 19, 20, 30, 32, 34, 36
Triazo-, 67
TRITC, 70
Triton X-100, 74
Tryaryl methanes, 66
Tube length, 46
Tungsten
  film, 52
  lamp, 52
Tungsten–halogen, 45
TV camera, 50

Ultra-long working distance, 46
Ultra-violet, 31, 33, 46

ULWD – *see* Ultra-long working
  distance
Urea–formaldehyde resin, 59

Vernier scale, 50
Vertical section (confocal), 3, 25
Viability, 65
Viable cell, 60
Vibrating knife microtome, 59
Vital stain, 60, 66

Wavefront, 17
Wavelength, 5
  fluorescence microscopy, 21, 23

Wax, 59
Wide-field, 47
Wollaston prism, 16, 17, 18
Working distance, 46
Working temperature, 44

Xanthenes, 66
XBO lamp, 43, 45
Xenon, high-pressure arc lamp, 43, 45
XRITC, 70
Xylene, 90

z axis, 25
Zernike, 10
Zone of least confusion, 41

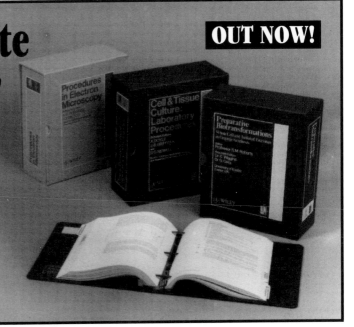

ESSENTIAL DATA SERIES

All researchers need rapid access to data on a daily basis. The *Essential Data* series provides this core information in convenient pocket-sized books. For each title, the data have been carefully chosen, checked and organized by an expert in the subject area. *Essential Data* books therefore provide the information that researchers need in the form in which they need it.

Centrifugation
D. Rickwood, T.C. Ford & J. Steensgaard
0 471 94271 5, March 1994, £12.95/$19.95

Gel Electrophoresis
D. Patel
0 471 94306 1, March 1994, £12.95/$19.95

Light Microscopy
C.P. Rubbi
0 471 94270 7, April 1994, £12.95/$19.95

Animal Cells: culture and media
D.C. Darling & S.J. Morgan
0 471 94300 2, due 1994, £12.95/$19.95

Enzymes in Molecular Biology
C.J. McDonald (Ed.)
0 471 94842 X, due 1994, £12.95/$19.95

Vectors
P. Gacesa & D. Ramji
0 471 94841 1, due 1994, £12.95/$19.95

Nucleic Acid Hybridization
P. Gilmartin
0 471 95084 X, due 1994, £12.95/$19.95

Human Cytogenetics
D. Rooney & B. Czepulkowski (Eds)
0 471 95076 9, due 1994, £12.95/$19.95

ORDER FORM

Please send me:

Qty	Title	Price/copy	Total
....	....	....	....
....	....	....	....
....	....	....	....

All prices correct at time of going to press but subject to change. Your order will be processed without delay, please allow 21 days for delivery. We will refund your payment without question if you return any unwanted book to us in resaleable condition within 30 days. All books are available from your bookseller.

Method of payment

☐ Payment £/$_____ enclosed (payable to John Wiley & Sons Ltd).
Orders for one book only – please add £3.00/$4.95 to cover postage and handling. Two or more books postage FREE.

☐ Purchase order enclosed

☐ Please send me an invoice
(£3.00/$4.95 will be added to cover postage and handling)

☐ Please charge my credit card account

    ☐ American Express   ☐ Diners Club
    ☐ Visa   ☐ Mastercard

Card no. |_|_|_|_|_|_|_|_|_|_|_|_|_| Expiry: |_|_|_|

Signature: _____

Telephone our Customer Services Dept with your cash or credit card order on 0243 829121 or dial FREE on 0800 243407 (UK only)

Send my order to:

Name (PLEASE PRINT) _____

Position: _____

Address: _____

_____

_____

_____

Telephone: _____

Signature: _____ Date: _____

Return to: Rebecca Harfield, John Wiley & Sons Ltd, Baffins Lane, Chichester, West Sussex PO19 1UD, UK. Telefax: (0243) 539132
or: Wiley-Liss, 605 Third Avenue, New York, NY 10158-0012, USA. Telefax: (212) 850-8888

☐ If you do not wish to receive mailings from other companies please tick this box or notify the Marketing Services Department at John Wiley & Sons Ltd.

Ⓦ WILEY